# Informational Nature of Being

*Road to Digital Divine*

# Informational Nature of Being

# Road to Digital Divine

## Computational Nature of Mind and Matter

———

*Seeking Eternal Truth and Wisdom Series*

## Hemant Gupta

**To order additional copies of this book, contact:**
Xlibris Corporation
1-888-795-4274
www.Xlibris.com
Orders@Xlibris.com
79861

# Contents

*To My Mother*

# Acknowledgments

A number of friends and my family members have helped me greatly in writing, editing, and inspiring me to write various contents through their stories. First and foremost I must thank my wife, Seema, and children, Vinay and Monish, for their love, support, help, and inspiration. Besides being my partners in crime of sharing this existence, I must acknowledge that all have been incredible sources of learning for me that has enabled me to understand the nature of my own self and the universe that I share with them. I am truly indebted to the incredible wisdom and love they bring to our existence. I owe deep gratitude to my father, Dr. Braj Kishore, who not only helped throughout with the process of editing but also provided invaluable wisdom through our weekly chats for many years now. I must also thank Raj and Grace for the many, many years of friendship and love. I am also deeply indebted to my younger brother, Shishir, whose marvelous poetry on self inspired me to write this series of books. I am very grateful to my sister Dr. R. Singhal for her support and guidance throughout my life. I would like to acknowledge Al and Ellen Fletcher, and Teresa for providing me over the years, the meaning of what is being a true saint. I would also like to thank Jona for sharing her wisdom and spirit to live every day with incredible vigor and liveliness. Last but not least, I would like to thank Ruby for allowing me to share her story and inspiring all of us with the courage she has shown and to see what a great heart she has.

—⁓⁓∘●⊱⊙●⊰∘●⁓⁓—

# Computational Universe

The universe is made of bits. Every molecule, atom, and elementary particle registers bits of information. Every interaction between those pieces of the universe processes that information by altering those bits. That is, the universe computes, and because the universe is governed by the laws of quantum mechanics, it computes in an intrinsically quantum-mechanical fashion; its bits are quantum bits. The history of the universe is, in effect, a huge and ongoing quantum computation. The universe is a quantum computer.

—**Seth Lloyd** (*Programming the Universe*)

> **Computation is not a machine process . . .
> [It] is an abstract mathematical process that
> exists only relative to conscious observers.**
>
> —Searle

# Chapter 1

# Road to Digital Divine

We are Divine enough to ask and we
are important enough to receive.

—Wayne Dyer

# Introduction

"Suppose you were timeless."

"What do you mean?"

"Means you cannot die."

"Okay."

"Suppose you have anything and everything that you ever wanted or desired!"

"Really?"

"Yeah, any thing you wanted or needed."

"Really? Okay."

"You were infinite."

"What do you mean?"

"Means you are everywhere."

"That's odd! But okay."

"You were perfect."

"You mean never made any mistake?"

"Yeah."

"Okay."

"Then what would you be doing now?"

Baffled, trying to fulfill his obligation to answer these weird questions from his dad, eight o'clock in the morning, and trying hard to keep his mind away from his digital acquisitions, my thirteen-year-old Monish attempted to force this reply out of his reluctant mind, "Sounds like you are once again confused about God, Papa! Well, I would do what God does! Just watch and help, as you know 'run this universe.'" He pretended to sound interested and meaningful.

As I dropped him to his school and headed to my work, his words reverberated in my mind. "Why is a thirteen-year-old so clear about God that has me confused and unclear about its existence and nature for so long?" I pondered. At one time, earlier in my younger days of learning and practicing hard-core science, I had never allowed my mind to entertain such thoughts about behavior or even existence of such a godlike entity. However, today, I must confess, it has become my obsession.

With incredible progress in science over the last century, the inner workings or our universe is becoming clear both at the physical and computational level. This is leading to an incredible paradigm shift. It is leading us to see universe as a grand, digital, informational, computational entity. Hustling and bustling with incredible dance of bits, bytes, and quantum bits, creating incredible array of events and action, almost like a living entity. Is there a God, divinity, or even a digital divine buried in the nature of this entity?

Most of us are conditioned to think that we are all distinct three-dimensional beings and live in a physical universe comprising of three-dimensional space-time where time represents change. As the gap between science and spirituality closes, a new dimension of our perception is emerging that is challenging this view. Information is at the root of it all. The words like *data, knowledge, wisdom, programs, meaning,* and *intelligence* reflect the mode of inner workings of this universe. The concepts like systems, emergent properties, entanglement, oneness, information processing, mind, and self-knowledge become the hallmark of this paradigm. All the complexity of mathematics, physics, chemistry, biology, and even spirituality all reduce into a simple computational reality reflected by posing of simple yes/no answers.

This is so bizarre and hard to believe and yet forms the cornerstone of our new reality. A reality that undermines the separateness of all, and highlights oneness, that resides at the root of all existence. The oneness that is critical for us to understand, benefit, and manipulate our true reality. It is the dimension which arises out of advances in mathematics or science of computation or science of information. It is the information processing which transcends space-time, acting in concert with our physical universe, which runs in a computerlike manner. Our universe is a mega quantum computer that is performing a grand act of information processing. What is it computing? It computes itself and its components, says Professor Seth Lloyd of MIT, in his book *Programming the Universe*, a pioneer in the field of quantum computing. In addition, there appears to be a seamless link between the separates in physical dimensions and oneness of this computational dimension.

This feature of computation has remarkable likenesses to the dimensions of spirituality. So-called mystics may now believe that finally there is an expression of science, which supports what they have always believed in. A dimension that is free from constraints of space and time, that is timeless, and spaceless. A dimension with computationally executable programs, like computer programs we are familiar with or instruction sets of a special kind. A worldview in which all interactions in a space-time world are rooted in computation. A perspective that involves computational elements of space-time floating and interacting with one another in various mathematical forms such as functions, equations, and computational systems. Some may even call it the dimension of divinity, as its description resembles closely to omniscient, omnipresent, and omnipotent description of God. A dimension of miracles, greatness, and grandeur as it succeeds in creating and managing infinity of the universe. It is a dimension of perfection, eternity, and infinity. A dimension where all dualities merge, and one emerges! A dimension of, one may say, unconditional love, an essential ingredient of oneness. A depth in which supreme design, connectivity, and management live. An incredible capacity, which represents supreme intelligence and perhaps supreme empathy. Undoubtedly, a dimension of universal love, or connectivity, with omniscient, omnipresence, and omnipotence characteristics.

Humans have intuitively felt the computational nature of our existence since ancient times. Scientific methods involving theory, control experimentation, observation, and analysis evolved from that belief. Newtonian physics almost came to a conclusion that universe, including us, are a part of a physical, mechanical device that works with complete determinism, reductionism obeying mathematical theories underlying the Newtonian physics. This was a different universe that spiritual and religious belief led to, where a holistic and empathetic universe with divine, unconditional love was the central principles of universal existence.

An even deeper computational nature of existence has evolved from the advances in modern science. Latest scientific studies have shown that this deeper computational dimension lies at the root of our existence and goes well beyond what Newtonian physics could realize. It is the dimension of digital divine. Later I will explain the relevance of this emerging dimension with our ever-growing understanding of a quantum world and evolving principles of quantum physics. From this perspective, the universe still appears to be a machine, but a different machine than Newtonian physicists suspected. A quantum universal information-processing system that is holistic, empathetic, and exudes promise of unconditional love, just as spiritual wisdom projected. Same to be true for life including human beings, which can be viewed as quantum information-processing entities, linked with universal quantum information processor, connected to one another using local as well as nonlocal quantum communication channels. At the least, the gap between the scientific and spiritual wisdom is narrowing in a meaningful way.

I have always been an ardent advocate of systems view of the universe. A repeated theme that you will encounter will be that the universe is one system, and all in it are interactive subsystems. Over a period, we have developed vocabulary such as *matter, mass, particle, waves, energy, silence, machine,* and *devices* to describe this one system, our universe. I will start with a general discussion on the devices of nature operating between silence and energetic states. A deeper discussion on the nature of *silence* and *energy* will reveal a computational nature of universe and its devices. From this point of view, the silence will appear, not as profound emptiness, but more like grand fullness or an incredible inkpad on which energy, the ink of creation, has written the equations of universe. We will explore Newtonian determinism with quantum uncertainty and general nature of quantum entities leading to atomic, subatomic, and molecular entities to further our grasp of these fascinating devices of nature. We would learn to recognize the visible universe as the interactive collection of quantum mechanical, computational devices operating between silence and energetic states. More importantly, we will develop a model of our universe as a giant quantum computing system where a quantum bit or qubit is the smallest unit for processing quantum information, and the universe itself is a grand quantum information-processing and distribution system. This computational system can also be described as a cosmic mind, just to emphasize its similarities with the human mind. How does life arise through a process of self-assembly, from this quantum distribution system, and masters a binary information generation and distribution system, leading to modern humans and human society? I will address how the quantum and binary information switches need intelligence or mind to make cooperative decisions.

Today, we are forced to examine our reality based on at least three levels of existence. First, is the physical level. It is the physical universe that we see, touch, or feel. I have portrayed this physical universe as myriad of interactive systems or one may say the system of devices. From this view, humans are one such device of the physical universe. The second reality is mental reality, which is a property of a conscious observer that is able and desirous to make sense out of this existence. The mental reality leads us to another reality, which may be the most absolute reality of the three. It is the mathematical or informational reality. I would refer to this reality as computational reality stemming from its likeness to computers or more precisely computational algorithms. There is another, fourth level of reality, which may be even more critical to access true reality. It is experiential level. It relies on subjective experience of an observer or observer-based reality.

With advances in quantum physics, the observer has become central to experiencing and creating any sense of reality. Our senses confirm that we are separate from the rest. Nevertheless, the connectedness at the quantum level as one defies this description of reality. How do limitations of our senses affect our view of the true nature of reality? What kind of perspective leads us to *unity consciousness* as suggested by quantum nature of our universe? Does this nature of universe turn our concept of reality upside down? What is the true nature of reality? As it will become clear that much of human sufferings root from an old perspective of separation-based binary thinking, inculcated deep in our psyche, especially reinforced during the evolution of materialism or an era when Newtonian deterministic belief reigned supreme. A newer mode of human thinking, rooted in quantum nature of our being, will revolutionize the way humans look at their world, their challenges, and their joy. Once understood well, the world will never be the same and collective human sufferings will gradually disappear. In order to accomplish this, we must take that road to understand the new form of divinity or better yet digital divine.

———〰⦿❧⦿❧⦿〰———

# Nature of our Physical Universe

# Chapter 2

# Systems of Nature

# Operating Between
# Silence
# and
# Energetic States

Classical physicists considered machine as any device that used energy to perform some motion or activity. The term *device,* according to *Webster* dictionary, is a piece of equipment or a mechanism or an entity designed to serve a special purpose or perform a special function. "My physical universe is what I see, sense, and feel." This was the perceptual reality that led classical physicists to understand our universe as a large machine or a system of machines or devices comprising of many interactive smaller machines. In other words, the universe to these physicists was very much like a mechanical clock of their times, constantly running and keeping time with a series of interacting physical or mechanical gear system. A paradigm, which is not quite wrong as it is incomplete. All outcomes in Newtonian world were deterministic and firm. Once the first conditions were set in motion, there was no room for a change. A big surprise came when in this description there was no role for God. The equations describing the universe had no variables that only godlike entity would satisfy. The mathematics worked without any divine intervention.

Today, our modern worldview is rooted in information or information systems, systems of machines or devices comprising of matter and energy, all processing information. Information is at the root of all matter and energy. This allows our universal reality to have much deeper and broader meaning, especially as one incorporates systems capable of information processing at the quantum level. In this description of universe, the mathematical equations emerge to be such, that the variables they contain do need, which some might call, godlike entity to satisfy them. This is important, but not necessary, to harmonize science with spirituality or religion.

Let us start by understanding what is a system. A system, according to Austrian biologist Ludwig von Bertalanffy: "A system is an entity which maintains its existence through the mutual interaction of its parts."

Emphasis here is on "mutual interaction," something that takes place between the parts, over a period of time, which is essential to preserve the system. Matter or energy is the most general description of part of a physical system. Matter is a physical body, a physical substance, or simply a form that occupies space and can be perceived by one or more senses. Information systems include information as the most basic part. Quantum systems include quantum particles or information as interactive parts of these systems. In general, systems could involve matter, energy, and information as interactive components.

A system implies something beyond random cause and effect. This is a crucial point to understand. Systems can be quantum particles, atoms, molecules, cells, organs, persons, community, state, nation, world, solar system, galaxy, and the universe itself, in increasing levels of complexity. All devices or machine can be called systems as well. The fact is that there is only one system, *the universe*. Every other system, device, machine, and myriad of names given to all that exists in this universe is really just subsystems of this larger system. Where one should choose to draw the boundaries or where the boundaries are already drawn by nature is the key question. An observer is central to defining the system, system of systems, subsystems, and their parts.

Systems may be either closed or open. A closed system does not have to interact with its environment to maintain its existence. Mechanical systems are classically considered closed systems. In older times of classical physics, atoms and molecules would be good examples of such systems. Open systems interact with their environment in order to survive or maintain their existence. Humans are open systems in that they must interact with their environment in order to take in food, water, and obtain shelter. Humans release waste products to the environment in return. An open system may interact with its environment and grow, shrink, or establish harmony by striking a balance with the environment. Classical physicists considered systems interaction with the environment to be limited to material or energy. Today, information is at the root of all interactions. All systems that fundamentally exchange information with their surroundings are considered open. With our growing understanding of quantum physics and principles of entanglement, the traditionally considered closed systems may be actually open systems. From a perspective of quantum information, all subsystems are open. There may, in fact, be no closed system except one, i.e., our universe.

Why do we worry about boundaries within a system or defining subsystems within a system? The intent of investigating systems, in the form of subsystems, is to develop an understanding. In fact, taking the larger system into account is often necessary for understanding the operation of the subsystem. It is often more appropriate or necessary to consider a system not only as a subsystem of some larger system, with which it must interact in some way, but also as larger system to all smaller systems it contains, in order to develop a complete operational understanding. Since, for our universe, despite of the latest advances in our scientific understanding, we have not been able to define boundaries of either the largest or the smallest system, it means all systems are both subsystems of larger systems and consist of subsystems at the same time.

## The View, Bottom-Up vs. Top-Down

Descartes and Bacon provided us with an analytic framework for developing understanding using the scientific methods. The Newtonian paradigm embodies systems comprising of large, visible, solid classical objects that essentially hold a linear cause-and-effect relationship. This view is reinforced by the way our senses perceive our physical world. It, therefore, continues to be a significant top-down view of our reality. Our new understanding of deeper reality at the quantum level attempts to complement our reality models from a bottom-up view. The bottom-up view is one of the most noteworthy perspectives one can have. Most of that is natural happens through a bottom-up approach. Larger structures or systems evolve from smaller beginnings. Both top-down and bottom-up views are useful; however, there are serious challenges at the interface, where the two views meet.

A systems approach provides us direct insight into these challenges. It implies that the foundation for understanding lies in interpreting interrelationships within systems and in between systems utilizing both top-down as well as bottom-up views. At the interface, the results should harmonize. Connectedness, entanglements, or interrelationships within a system and between systems is responsible for the manner in which systems operate as well as result in the patterns of behavior and events we perceive. This is particularly critical for quantum systems that connect and are naturally entangled with one another. Since our universe is one system, harmonizing top-down and bottom-up view of all systems it contains can only be done by true understanding. Until one arrives at such truth, the search continues.

Often we use the words *device* or an *entity* to describe a system in different contexts. The words *device* or *entity* are often used as a special top-down description of a system as a whole. The parts that make the whole or an entity or a device reflect a bottom-up point of view. The parts or subsystems within a system in turn may also be looked at as systems of smaller components that make them or devices depending on the intent of description to be top-down or bottom-up.

The top-down view is extremely familiar to us humans. This, in part, has been responsible for great progress that human society has made. We conceptualize, design, make blue prints, and manufacture the devices and systems. This is top-down thinking. The top-down approach is like a CEO's perspective. The executive order that flows down. Externally visible features, data through our observations, are the hallmark of this reality.

The nature, on the other hand, works from the bottom up. The nature does not care to share with us its true inner workings. This is something humans had to infer by observation, experimentation, and analysis stemming from top-down reality. We often have to go beyond the linear cause and effect patterns of behavior to infer the systemic interrelationships among the parts of systems in order to develop a much deeper understanding of the nature. What we are attempting to put together is a bottom-up view of nature. This mental, top-down reductionist approach has been incredibly successful and may ultimately lead to a true bottom-up world perspective through rational, scientific analysis.

In fact, any system can be observed by either a top-down view or bottom-up view. Exception may be the universe itself. Since human observer has not devised a way to step out of the universe it resides in, and hence it can only observe the universe as a whole from only within. There is, however, another way to access this reality, at least for human observers; and it is through feelings, which are the route most mystics, religious, and spiritual followers take. Science has stayed away from this method of investigation as results have not cleared the reproducibility and objectivity standards demanded by scientific methods. Clearly outcome is subjective in nature. These two perspectives form one of the most crucial concepts in understanding ourselves, the concept of me, *myself*, and my true nature. The top-down perspective leads to my outer world, whereas the bottom-up perspective leads to the concept of my inner world.

As I look myself in the mirror, I see myself, and I can feel myself breathing. I'm inhaling and breathing out air. I am an entity from a top-down perspective. As a single entity, I could be an employee working in an organization with specific responsibilities. I am one among many other employees. What is critical here is how I function as a whole, my skills, education level, talents, and other utilities to the organization.

From a bottom-up perspective, I am a complex system made up of many smaller components or devices or subsystems. Inside my body, there are many structures or subsystems such as the digestive system, respiratory system, and ultimately about 50 trillion of living cells that make my body. I can configure the state of these cells or every molecule, atom, or quantum entity that is part of me at will with a top-down command. Or I may allow the configuration that is evolving from the bottom up, i.e., quantum level, to ultimately configure my muscle or other cells and eventually myself.

This is as if we are talking about two separate selves, one that could be configured from top-down perspective, my outer self, and the other one, let us say, my inner self that is configured from the bottom-up perspective. This is a truly critical know-how to understand the true nature of ourselves. We briefly talked about it in the last chapter, and I will come back to it again in the last chapter of the second book in this series as understanding our true self is the key to accessing eternal happiness or bliss.

## Emergence

An idea or principle called emergence is closely tied with the concept of system. It is often described in terms of emergent properties. These are the properties that arise from the mutual interaction of the parts of a system. However, they are not found as characteristic of any of the individual parts or mere collections of parts within the system. This means that the top-down description of a system, a device, or an entity is more than the parts of subsystems that make up the device or entity. A good example to understand emergent properties is table salt or sodium chloride. One could study sodium and chlorine in isolation from each other forever and never discover the key attribute of sodium chloride, i.e., its salty taste. Its distinctive taste is an emergent characteristic of the mutual interaction of sodium and chlorine when combined to produce the molecular form called sodium chloride.

One has to study the system to get a true understanding of taste of sodium chloride. Studying the parts will not provide an appropriate understanding. Often we can experimentally characterize as well as theoretically predict the emergent properties of a system or systems of subsystems. The nature and the domain of the system are derived from careful experimentations and understanding of emergent properties, often through a process involving scientific methods, mathematical modeling, as well as trial and error. When the two match, the understanding emerges, implying that the domain and the boundaries of consideration are likely to be appropriate. This is the engine that derives our modern scientific knowledge.

In a similar manner, energy manifests itself in the form of the quantum particles, quantum particles combine to form atoms, atoms combine to form molecules, molecules combine to form nanostructures, and nanostructures lead to macrostructures. At each level, novel emergent properties may be realized, greater the complexity of organization, greater the richness in emergent characteristics. A living cell is a system that shows emergent properties owing to being a system of smaller systems or devices. These properties are frequently used to assess similarities or differences between living cells. For example, an amoeba, paramecium, and a nerve cell in my brain are decidedly different in terms of their nature and functionality but are similar in terms of mechanism of energy consumption or reproduction. Human systems also reveal emergent properties that make us unique and more than simply the collection of trillions and trillions of cells. That I like a game of poker, love to watch TV, and generally love to sleep are the emergent property of trillions of cells that constitute me. I doubt if any of these are shared by the cells that make me as an entity. But a complete description of me cannot simply cater to me as an emergent entity. Trillions of cell that make me, acting as independent and communal contributors, complex molecular chemistry involving atoms, and quantum entities are all part of me and form the bottom-up perspective that makes the informational entity that I truly am. Our universe is also a system and a collection of many subsystems. What are the emergent properties of our universe?

A systems view is holistic and may appear somewhat in contradiction to the concept of scientific analysis based on reductionism, which is breaking things down into smaller pieces to simplify the study. Reductionist approach may potentially lose the relevant characteristics of the system or possibly lead to development of a less-than-complete understanding. The study of the system, however, involves the whole and the parts, aiming to develop a complete understanding. Both bottom-up and top-down views are essential.

Without a doubt, systems approach has provided us with many answers, but it leaves us with many questions as well. What is the place of God in this universal systems approach? *Who* am I? What is my true nature? Am I a mere system or assembly of systems embedded in silence, entangled with other systems, and involved in some grand act of information processing? Is there a purpose of this assembly? Is there a designer? If there is one, who is it and what is the purpose? Is the assembly itself a grand entity comprising of infinity of devices operating between silence and energetic states? I leave you here with more questions than answers. This is how human quest to gain knowledge began. It led to the birth of science. The science turned out to be incredible weapon that would dominate human existence. Taking that bite of the apple of knowledge became inevitable. The Garden of Eden would never be the same.

# Chapter 3

# Nature of Silence

# Nothingness
## or
# Fullness

# Something from Nothing

Can something emerge from nothing? Alternatively, you may ask, can space-time arise from spaceless and timeless universe?

As hard it is to believe, nothing has produced all that we see around us!

Yes. Today's science implies that our visible universe has evolved from nothing.

It is contrary to our normal instincts. What about the laws of conservation of mass, energy, and information? You may ask. The answer may lie in the mystery of nothing! What is nothing? Is it spaceless, timeless, beyond space and time, zero, silence, or something else? As mysterious as nothingness is, it is all around us in our physical universe, in our mental world of understanding, and in our mathematical world of formulas and functions.

Hindus have been credited with the invention of zero, a number that was perhaps the first attempt to quantify nothingness. It turned out to be a number that changed human history. In fact, it will not be an exaggeration to describe modern era for humans as post-zero era. In a computational space, zero is easily defined and useful. Mathematically, zero is easy to understand. It is such an integral part of our modern lives that it is impossible to imagine our modern lives without the number zero. In our number system, zero is a count that means no value or quantity. If I have zero money, that means I have no money. It sounds easy. What does zero mean in a physical universe? It gets harder and harder to define. As the physical size of an object gets smaller, our ability to make the observation of its dimension, location, and momentum gets harder. As the timescale of an event gets smaller or vibration frequencies approach zero, once again our capacity to be able to make measurements diminishes. In a physical universe, zero space-time, vacuum, and zero molecular motion all represent physical barriers, all of which seem unattainable. In thermodynamics, zero becomes uncrossable barrier, the coldest possible temperature.

For matter, the journey down the temperature scale is a story that involves losing energy and increasing internal order. As one cools the matter, it becomes quieter, condenses or solidifies. For example, the molecules in liquid water are free to move around in the liquid phase but freeze or organize in ordered crystal lattice when liquid water turns in solid icy state. At lower temperatures, even more ordered states of matter evolve. The superconductivity and superfluidity are two of the most observed states. These states are possible as atomic particles coordinate their motions at quantum levels. The coordination at this level results in extraordinary properties. For example, electric current flows with little or no resistance in superconductive solids. In superfluids, liquid flows up, defying gravity. At root, these behaviors are the result of groups of particles at the quantum level behaving as *one*.

The holy grail of ultralow temperature physics is the absolute zero temperature. Absolute zero itself is unreachable. We can get infinitely close to it but can never reach. It would take a cooling machine of the size of our universe and infinite time to reach absolute zero. Even if we get there, some energy will stay in the system. It is known as "zero point energy." The uncertainty built into quantum mechanical behavior of the universe would ensure this infinite, omnipresent energy exists, even in the deep vacuum, and results in a force exerted by nothing at all!

Very close to this state, groups of atoms will condense into a single entity known as Bose-Einstein condensate, named after two physicists Satyendra Nath Bose and Albert Einstein, who theoretically predicted this exotic state of matter in 1924. At these low temperatures, the wave function of many particles merge into one wave function and the group of many particles behave as one entity.

The closer one gets to absolute zero, the harder it is to get any colder. The atoms tend to grab pockets of energy from just about anything in their immediate environment. It is only reasonable, for most of us, to expect the coldest spot in the universe to be somewhere in the vast emptiness of space between galaxies. This is perhaps not true. Scientists are within a millionth of a degree of absolute zero, closing in this unreachable thermodynamic barrier. The coldest spot in the universe lives very likely here on this earth, perhaps in the laboratory of a mad cold temperature physicist.

What happens when we heat up the states of matter? Yes, quite opposite to what happens to the states of matter when we journeyed down the temperature scale. Instead of removing energy, we are adding energy to the system. The matter states of decreasing order result, as it journeys up the temperature scale, or more energy is added to the system. Let us examine what happens when we heat up ordinary water. It, first, changes to vapor, a gaseous state. In comparison to the liquid state, the water molecules have greater freedom to move about in gaseous vapor state. If we heat it even more, the water molecules dissociate into oxygen and hydrogen atoms. A life-giving liquid, wet and inert, is now two highly reactive gases.

If we heat this atomic soup some more, the atoms no longer exist; they split into smaller constituents, the electrons and the nuclei. This mixture is called plasma. It is electrically charged, and under the right conditions, it would glow. Of course, one can continue in a particle accelerator and split the nuclei into protons and neutrons. Depending on how far one is able to go, one can, perhaps, disassemble the protons and neutrons into even smaller constituents called quarks. According to some, disconnected quarks may form the matter at the very top of the temperature scale. It has been named quark matter, a sea of quantum entities that are so weird that no one has been able to estimate their properties.

The electroweak unification has been demonstrated at a temperature equivalent to $10^{15}$ K. The wavelength of the equivalent particle would be $10^{-16}$ cm. To probe, at shorter and shorter distances, we need even higher energy per particle, since distance probed depends on the wavelength of the particle, which depends inversely on its energy. At distances of $10^{-30}$ cm, the strength of electroweak and strong nuclear forces becomes equal. Finally, at Planck's dimensions of $10^{-46}$ cm the gravitation force becomes equal to other three fundamental forces. At the dimensions of Planck's scale or beyond, space-time does not exist, and a unified field is present everywhere in the universe.

In quantum theory, all quantum units are dimensionless points. For example, height, depth, and width of a quantum entity will be assigned only three numbers. By contrast, if you consider any physical object with its own height, width, and depth, you would have to assign coordinates to describe each edge of the object. In other words, you would need more than three numbers to describe the object. Quantum "particles" are designated as simple points, without size and thus without edges. There seems to be no clear way for physicists to determine coordinates for their outer edges or to determine if, in fact, they even have any outer edges. The three coordinate numbers are assumed to be sufficient to locate them as a single point in space. If precise quantum calculations are carried out all the way down to an actual zero particle size that is zero height, zero width, and zero depths, the quantum equations return meaningless results. As the matter is squeezed into a dimensionless zero point state, as described in Einstein's theory of general relativity, a zero becomes a black hole further intensifying the mystery of nothingness or emptiness. What is emptiness then?

## Emptiness

The empty space is actually not empty at all. It is a state that reflects infinite, dynamic quantum fluctuations and computation. A mathematical theory of such grand unified state is still to be worked out; however, it is becoming clear that it is a state that even though beyond space and time, it is teaming with communication signals. It is a state of existence where all dualities merge into one and a state of being where signals travel beyond the locality, at potentially infinite speed. Such quantum physical nature of nonlocal action suggests that information can be exchanged without any energy exchange. It is a form of inaction which is beyond action or energy. It is the inaction that also leads to silence, a state of pure observation and awareness, a stillness that surrounds all there is.

In Buddhism, emptiness forms the core of its philosophy as described in the ontology of Mahayana Buddhism. The maxim "form is emptiness; emptiness is form" is perhaps the supreme mantra, the most illustrious duality, linked with Buddhist way of life. The oldest of the Mahayana texts perhaps originated in India, around the time of Jesus Christ, and describes the emptiness as the key concept to understand its foundational values.

The Buddhist notion of emptiness or Hindu term *sunya* or zero is often misunderstood as nihilism. According to nihilism, emptiness means that nothing exists or reality based on emptiness is unknowable. It will have nothing meaningful to say about our reality and the universe. The Buddhist notion of emptiness, on the other hand, arrives at just the opposite. It claims that ultimate reality is knowable. We can communicate with emptiness and receive useful guidance, knowledge, and wisdom from it about our own reality and the universe around us.

How is that possible? Let us further try to understand the Buddhist meaning of this term with an example of an empty cup. We often say a cup to be empty if it does not contain any liquid or solid. How is this emptiness different from the Buddhist meaning of emptiness described earlier? Of course, it is the ordinary meaning of emptiness. But is the cup really empty?

You may say that it is really not empty as it is still full of air. Can a cup be empty of all substances? Imagine a cup in deep space surrounded with vacuum. It is clear that it does not contain any air, but it may still contain space it occupies. Hence, one may conclude that the cup may never be empty and may always be full of something.

However, from the Buddhist point of view, the cup may be empty. Not only that, a cup full of liquid may also be empty. How is it possible? It is because the Buddhist understanding of emptiness differs from the physical meaning of emptiness. In Buddhism, we need to be precise and state what the cup is empty of. If the cup is empty, in Buddhism, it means that it is devoid of its separate self or existence.

In Buddhism, emptiness, silence, or nothingness all represent a way one looks at what is experienced. It is ones mode of perception. It is called empty if it adds nothing or takes away nothing. This means acceptances of the raw data, of physical and mental events, and the way it truly is; no thoughts or interpretation are added. This mode of witnessing the reality is called emptiness because it is empty or devoid of the presuppositions we usually attach or add to experience, sometimes simply to make sense of it.

We all have experienced silence as a contrast to *noise*. "Please, do not make that noise. I would like some peace and quiet." Does it describe the nature of silence adequately? Is there a deeper nature of silence, emptiness, or zero state of being? What is the true nature of silence? This is where our scientific understanding falls short. It is also where another understanding begins. Throughout our human history, silence has played a critical role in almost all spiritual practices.

Are we aware of the silent observer within us? What happens when we are truly silent? What happens when, within my body, all my life systems scramble to achieve a natural state of silence while I am being alive and silently aware? We are aware of silence in spiritual practices like meditations or Samadhi. Is it the goal of these practices to arrive at a state of stillness where only energyless or actionless information exists and can be felt? Can one be still or perfectly silent and experience information exchange?

Is silence a potential grand unifying physical state of the universe? Is silence a state where all dualities merge? According to some, it is in this mystery of nothingness where our search truly begins.

—⸎⸎⸎⸎—

# Chapter 4

# Energetic

# Quantum, Subatomic, Atomic, Molecular, Nano, Macro, and Biological Entities

## or

## Simply a Grand Act of Information Processing

# Distances in the Universe

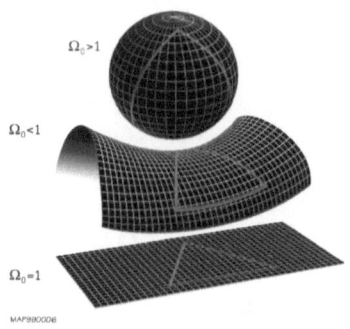

Distances in the Universe: Universe is the largest entity known ($> 10^{26}$ m).

Distances in the Universe: Cluster of galaxies could be as large as $10^{24}$ m.

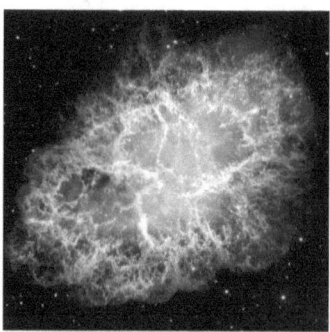

Distances in the Universe: A typical nebula is about $10^{16}$ m long.

Distances in the Universe: Our solar system is about $10^{13}$ m wide.

Over last few centuries, humans have been preoccupied with technological progress. This has impacted just about every facet of our lives. The engine, which has driven this progress, is based on advancements in scientific knowledge. This also has contributed immensely to our understanding of inner workings of our universe. More significantly, scientific knowledge has systematically dispelled myths, orthodox belief that existed in many of the world cultures and traditions. Over the last few centuries of scientific scrutiny, it is becoming clear that all the physical systems and devices of this universe are made up of two ingredients—silence and energy. As discussed in the last chapter, a state of silence or zero surrounds all that is energetic and part of our observable universe. These two ingredients together seem to form a cosmic blueprint for everything. The energy manifests itself in a variety of forms such as a matter or information. Today, our understanding is rooted in informational and computational view of our universe. How has scientific scrutiny led to evolution of computational or informational view of our universe? It is an incredible story. It is the journey that science took to learn deeper nature of matter, energy, and information.

## Let There Be Light, Ink of Creation

"Let there be light," according to biblical creation marked the onset of God's magnificent creation. It may not be too far from scientific views of the way this universe began its existential journey. Energy created this universe, at least the visible one. In fact, most widely held belief in science is that our observable universe began with a big bang. A huge explosion that resulted in spreading a vast amount of energy throughout the space. Over the period of time, the space-time expanded, the energy cooled, and resulted in quantum particles, subatomic particles, atomic particles, and myriads of complex matter particles which eventually led to life and humans.

What is energy after all? Most of us have some ideas about this word called *energy*. We use it in everyday language. Energy drinks, heat energy, light energy, chemical energy, nuclear energy. It is a familiar idea, and we all seem to know something about it. We know energy transforms from one form to another. We run, walk, and consume energy. We spend energy even in sleep. Everywhere we see there is energy. We know plants harness Sun's energy and begin a complex web of food cycle, which ultimately provides food to all of us. We know all food we eat is actually a form of energy. Many of us are

Distances in the Universe: Our Earth is about $10^7$ m in diameter.

Distances in the Universe: Wall of China is $10^6$ m long.

Distances in the Universe: A typical building is about $10^2$ m high.

Distances in the Universe: Coins are typically about $10^{-2}$ m in diameter.

familiar with Einstein's famous equation $E=mc^2$. It is so famous that we see it printed on T-shirts. It tells us that the mass and energy are interchangeable. All mass we see is a form of energy. All light we see is a form of energy. All electricity we use and all devices we use are just nothing but energy. However, do we know what energy truly is?

## Us and Visible World of Matter and Energy

Human fascination to understand the matter and energy around us goes as far as our collective memory. Today, human success as a surviving species owes much to our obsession to understand and manipulate matter and energy around us. Early humans learnt to distinguish between materials that were used for making clothes, shaping into tools, or suitable to eat. Their senses led them to differentiate between objects within objects, such as whole and parts, through boundaries that were clear and recognizable through senses of touch, sight, sound, feel, and taste. Empedocles, a Greek philosopher and scientist, proposed one of the first theories that took on to describe the world around us. Empedocles argued, "All matters are composed of four elements: fire, air, water, and earth." It was one of the first reductionist approaches to view or understand our universe. It formed the basis of the scientific exploration that followed such humble beginnings.

Our journey to gain scientific understanding of our universe has not been an easy one. Sheer sizes of objects along with events taking place over incredible distances and timescales have been mind-boggling. The discovery of some of these structures with mass, length, and timescales involved has been astounding, not to mention fascinating. Scientists had to come to terms with this incredible diversity. Nature has proven itself grand, no matter whether we looked at the tiniest level to the largest level.

Earth fascinated early humans. It provided the impression of solidity, stability, a center. It appeared to be flat, and everything in the heavenly sky seemed to have moved around it. This perceptual reality based on common sense was the dominant belief accepted in early seventeenth century. It was proven wrong. It was, however, not an easy road to gain acceptance of new wisdom that defied the perceptual common sense view. Galileo, in 1633, was condemned for asserting that the Earth revolved around the Sun.

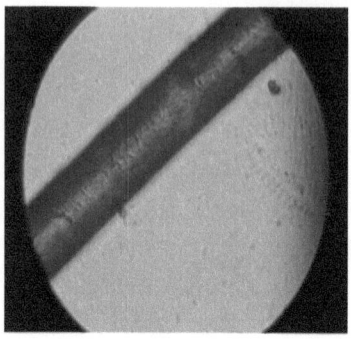

Distances in the Universe: Human hair is about $10^{-4}$ m in diameter.

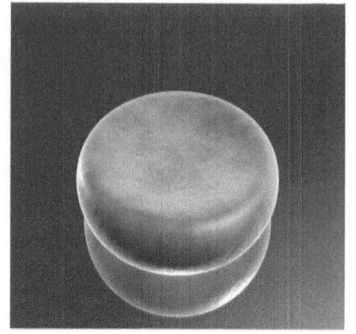

Distances in the Universe: A red blood cell is about $10^{-5}$ m in diameter.

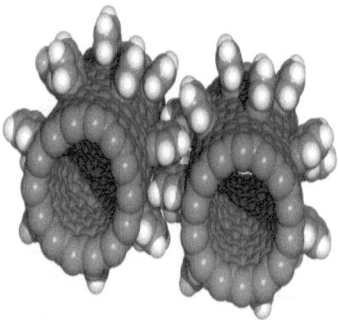

Distances in the Universe: Fullerene gear is a nanostructure about $10^{-8}$ m in diameter.

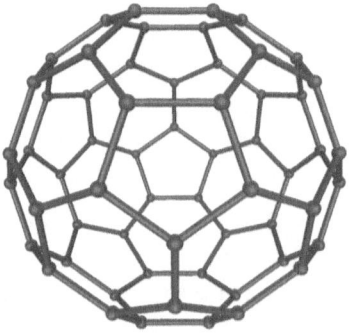

Distances in the Universe: Buckyball made from sixty carbon atoms is about $10^{-8}$ m in diameter.

The Inquisition, which sought out heresies, damned Galileo for backing a theory of astronomer Nicolaus Copernicus because it clashed with the Bible which said, "God fixed the Earth upon its foundation, not to be moved forever."

Today we know that Earth is just one of the many planets in our solar system that revolves around our Sun. And it is a sphere, not a flat object as our sensual perception implies. There is no doubt that the Earth is large and heavy. It contains $1.4 \times 10^{21}$ tons of matter, to be precise. It would make perfect rational sense to accept such a large object would be fixed or immobile. However, today we know better. We know that it moves, and not only does it move, it moves at incredibly fast speed of 67,000 miles per hour. No man-made object, at least the one that is visible to human eye, has been able to achieve that speed. In order to reach Earth orbit, a launch rocket must accelerate to a velocity of 17,500 miles per hour, which is roughly twenty-five times the speed of sound. Fast, but still nowhere close to the speed of Earth hurling around our Sun.

We also know now that our Earth is not the largest object in space. Not even close. Our Sun is bigger than Earth. To be precise, 333,000 times the mass of the Earth. Just a little bigger, at least from a cosmic point of view! Our Sun is a star. The stars are, in general, the most clearly visible objects in any galaxy. Each galaxy contains billions of stars.

Our galaxy, the Milky Way, is about 200 billion times the mass of the Sun. This means it is about 70 quadrillion times the mass of the Earth. Now that is huge! It is remarkable, but it is not the end. It is still only a small part of the universe. There are over 100 billion galaxies in our universe. Therefore, the saying that "our Earth is insignificant in the grand scheme of things" is not just an understatement—it is true!

Even though insignificant, relative to our perceptual universe, Earth is big. One trip around the Earth could earn us 24,901 frequent flyer miles. A superhuman flying a steady pace of 30 miles per hour, twenty-four hours a day, will require over a month to circumnavigate the globe. Of course, even superhuman flies at relatively slow speeds compared to the fastest speed known in our universe. The fastest known speed to us is the speed of light, which is about 186,300 miles per second. While our human flyer will take more than

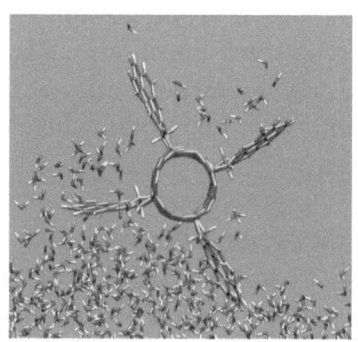

Distances in the Universe: A molecular propeller is about $10^{-9}$ m

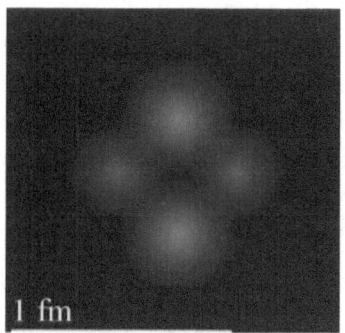

Distances in the Universe: Nucleus is about $10^{-14}$ m in diameter.

a month to get around the Earth, a beam of light would do the same nearly 7 ½ times in one second. Now that is fast! It is also the speed at which Einstein would often travel in his thought experiments. Just kidding!

In astronomy, we measure distances in light-year. One light-year is equal to about 5,875,156,800,000 miles. It is the distance a beam of light traverse in one year. This means going around the Earth nearly 236 million times! It would take our superhuman flier over 25 million years to travel across one light-year distance.

Our galaxy, Milky Way, is disk shaped. It contains about 400 billion stars. The disk spans about 90,000 light-years in diameter and roughly 10,000 light-years in height. All in all, it occupies a volume of over 63 trillion cubic light-years, an amazingly large structure! No wonder, until recently, we believed that our galaxy was the whole universe. But the reality is that it is only one among billions of other galaxies inhabiting our universe.

Just like each snowflake has a different crystal structure, all galaxies are different. The cosmic landscape is decorated by these incredible objects of incredible sizes and extraordinary shapes. These are spread out over incredible distances and appear like twinkling little stars, filling the night sky and us with incredible wonderment. Andromeda is the nearest neighboring galaxy to us, about 2.9 million light-years distance or thirty-two times the diameter of our entire galaxy! Even a beam of light will need 2.9 million years to bring that friendly greeting from Andromeda.

With powerful telescopes like Hubble, we can map even larger structures. Our Milky Way, Andromeda, and about twenty-eight other galaxies inhabiting a space about 10 million light-years in diameter are known as the local group. This local group is a part of a larger group known as the Virgo Cluster, which contains about two thousand galaxies spread over a space of 100 million light-years. The Virgo Cluster is just one among millions of clusters residing our universe. As far as we know, our universe spans distances more than 10 billion light-years that we can see. But there appears to be no end to it, and it may continue on to be infinite. Can there be a structure that is truly infinite?

Newtonian physics founded in seventeenth century is also known as classical physics. It was one of the first scientific attempts to understand incredible

# Time Scale in the Universe

Time Scale in the Universe: Creation of universe 1.4 × $10^{10}$ years

Time Scale in the Universe: Formation of galaxies 1.0 × $10^{10}$ years

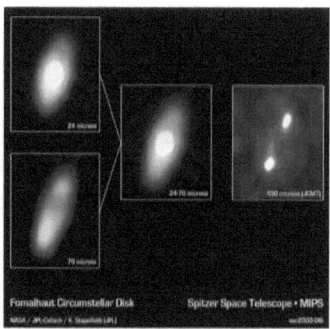

Time Scale in the Universe: Formation of solar system 4.6 × $10^{9}$ years

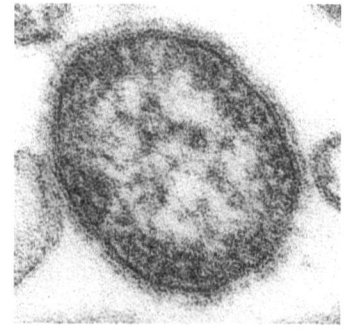

Time Scale in the Universe: Appearance of unicellular life 3 × $10^{9}$ years

array of distances, structures, and speeds of our universe. It assumed that the every thing in the universe comprise of solid objects. These objects are held together by a force called gravity. Newton's laws of motion described the motions of these solid objects including planets and other heavenly bodies with incredible precisions of positions and speed. These laws were so successful that, for a long time, we believed that these were the most fundamental, basic laws of nature. The classical physicists could not wait to penetrate the deeper reality of matter invisible to them. Little did they know that it would topple their worldview upside down.

## Invisible World of Matter

### Atomic, Molecular, and Nano Devices of Nature

If size was our guide to understand our visible universe, it became clear to classical physicists that there were objects that were invisible to the human eye, but let their presence known through bulk property measurements such as pressure, flow rates, or let us say, smell. In order to visually see these objects, the human eye would need to be aided with magnifying technologies. The nanoparticles are the largest of these invisible particles. One nanometer (nm) is one billionth, or $10^{-9}$, of a meter. If the size of a nanometer can be imagined to be like a marble, then a meter would be same as the size of the Earth. It is so small that is often hard to truly comprehend by most. One interesting and almost comical way I heard someone describe this scale is as the length, a man's beard grows in the time it takes him to raise the razor to his face.

A typical spacing between carbon-carbon bond lengths, in a molecule, is around 0.15 nm. A DNA double helix has a diameter around of about 2 nm. In comparison, even a single living cell is large. The smallest cellular life forms, the bacteria, are around 200 nm in length! A red blood cell is roughly 7,000 nm wide, and a water molecule is almost 0.3 nm across. Today there is an incredible interest in science and technology at nanoscale because it is, at this scale, the properties of materials can be significantly different from those at a larger scale. This is the scale where a bridge between classical and quantum behavior of matter can be observed. It also offers the prospect of using features from both worlds to produce exciting new behaviors and technology.

Time Scale in the Universe: Exotic creatures of Precambrian era up to $6 \times 10^8$ years

Time Scale in the Universe: Appearance of dinosaurs $2 \times 10^8$ years

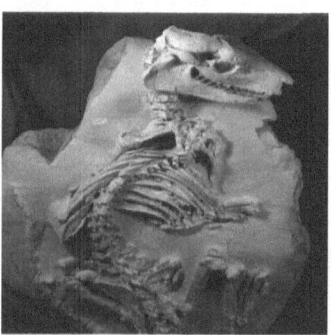

Time Scale in the Universe: Appearance of mammals $3 \times 10^6$ years

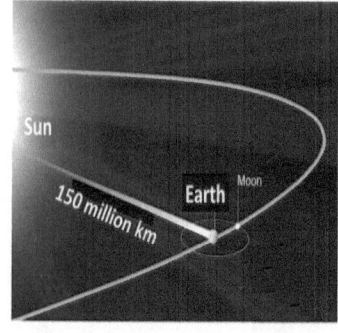

Time Scale in the Universe: Time it takes for Earth to move around Sun -1 year

Nanoscale devices can be designed to use the bizarre quantum behavior and classical features both working in one device. Life reflects one such design, and as we will see, it has resulted in a wide variety of living devices, humans being one of them.

All visible matter is made up of either molecules or atoms. Molecules are made up of atoms, so let us start with molecules first. How small are molecules compared to the small objects we can see with our eyes? Sitting on a sandy beach, we are often amazed at the number of sand grains around us. Such a large number is clearly not easy to count. The number of molecules of silica ($SiO_2$) in just one grain of sand could be even larger. Sand grains generally fall in a size range between 0.1 and 2 millimeters in diameter. If you made a few basic assumptions, you could estimate the number of molecules in average grain of sand. It turns out to be about 100,000,000,000,000,000—a large number. Therefore, a tiny grain of sand is enormous relative to a molecule!

Molecules are everywhere. From our body parts, tissues, blood, cells, food we eat, air we breath, water we drink, and to the house we live in, all are eventually made up of molecules. A molecule, in the simplest case, may have only two atoms, much like a tiny barbell with one atom at each end. Molecules could be more complex and involve a few to even several millions of atoms. These atoms could be arranged in all kinds of different shapes and sizes. The molecules like polymer are large and have been proven useful to human beings. For example, polyethylene bottles, containers that we use every day, are all made from polymer molecules, which could have well over several million of atoms in just one molecule! Our human body is made up of many, many large polymer molecules like proteins, carbohydrates, and fatty acids. Molecules represent the nature's mode of combining various smaller atomic level structures and creating larger organizational structure.

Just as humans could be part of a family, corporation, or political group, atoms can be part of larger molecular groups as well. Just as there are a large number and varieties of organizations that bind humans together such as corporations and families, there is a large number of varieties of molecules that bind various atoms together too.

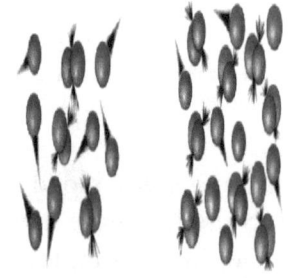

Low concentration = Few collisions    High concentration = More collisions

Time Scale in the Universe:
Time it takes for molecules
to collide—$10^{-14}$ seconds

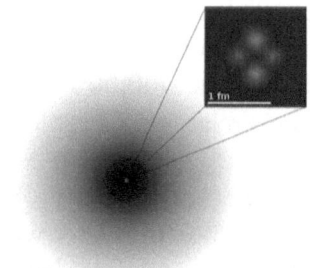

1 fm

100 000 fm $_{(= 1 \text{ Å})}$

Time Scale in the Universe:
Time it takes for light to cross
a helium atom—$10^{-17}$
seconds

Molecules are always in motion. Even a stationary object like chair has molecules that are in constant motion. The molecules can vibrate, with the barbell becoming alternately longer and shorter, and they can rotate, whirling end over end like a tiny baton. Large molecules like polymers reveal even more complex mode of motion such as reptation, a fancy word for snakelike motion. Molecules are constantly colliding with each other. Chemical reactions may take place during collisions when molecules hit one another randomly, sometimes end to end, sometimes in the middle. Under right circumstances, the chemical reactions lead to formation of chemical bonds. These chemical reactions are like human relationships. Sometimes we bump into someone and like that person so much that we form an instant bond; we become friends or lovers. No wonder, the word *bonding* is used frequently to describe such human relationships.

Can we observe the inner workings of a molecule? New twenty-first-century tools put us on the verge of this new science. We will soon not just photograph physical phenomena on the timescales relevant to atomic and molecular physics, chemistry, biology, and materials science; we might be able to develop animation clips combining ultra high-speed photography and quantum mechanical simulations. In nature, the relevant time markers span a truly broad range—from the dizzying attosecond ($10^{-18}$ second) timescale, the time it takes an electron to orbit an atom, to the timescale of $\approx 14$ billion years, the age of our visible universe. The relevant timescales for the atomic and molecular processes of interest cover nineteen orders of magnitude—a factor of 10 billion billion. The complex folding of a protein molecule can take milliseconds ($10^{-3}$ second) or longer , a quick reaction compared to human reflexes on one hand. However, on the other hand, a millisecond is an extraordinarily long time for an atomic collision, unless of course these atoms are supercooled close to absolute zero. The small molecules in the air we breathe undergo collisions with one another around every 100 picoseconds (ps) ($10^{-10}$ second), and they tumble or rotate in space about every 1 to 10 ps ($10^{-11}$ to $10^{-12}$ second).

The atoms within molecules do not lose their identity entirely. In fact, they behave as if they were balls joined to each other with springs. This, of course, is a gross simplification, and actual structure is enormously complex. Inside a molecule, a lot of dynamic action goes on. These balls and spring vibrate at a fast pace, with timescales involved in the range of 10 to 1,000 femtoseconds (fs) ($10^{-12}$ to $10^{-14}$ second), the spring representing the bond between the atoms.

Much of what we know about atoms and molecules comes from observing motion, starting with the direct observation of the Brownian motion of dust colliding with air molecules, explained by Einstein in a famous paper published just a century ago. Scientists are now attempting to capture motion not just between molecules but also within them. This would provide a basic understanding of chemistry and biology at the single molecular level.

All molecules are made up of ninety-two different varieties of atoms discovered so far. If you thought molecules are small, atoms are even smaller. Hydrogen atom is the smallest atom known. Its diameter is only about 5 × $10^{-8}$ mm. If we stack 200 million hydrogen atoms side by side, it would make a line just 10 millimeters long.

What is inside a hydrogen atom? Nothing! It is mostly empty. The electron spins quite far from the nucleus and in between, nothing. If a proton had a diameter of 1 cm, the hydrogen atom's electron would spin at 500 meters from the nucleus. In other words, the hydrogen atom would be as large as a football field, and the tiny football, in the middle of the center line, would be nucleus! In between, nothing or emptiness.

We are well aware of the limitations of the size of objects we can see with our eyes. Microscopes enable us to view objects that are smaller, and we are not able to see clearly with our eyes. For example, a drop of pond water looks very clear to our eyes, but if you look under an optical microscope, you may find it filled with tiny microorganisms like amoebas. However, objects much smaller than one micron across cannot be resolved with such microscope. Individual atoms are about 2 angstrom across, far too small to be seen under an optical microscope. How can one see the atoms?

In the early 1980s, Gerd Binnig and Heinrich Rohrer invented scanning tunneling microscope (STM) at the IBM research laboratories in Zurich, Switzerland. The STM, capable of imaging the atoms with resolution of a fraction of an angstrom, allowed some of the most spectacular and detailed images of atoms ever made.

J. Dalton was the first to propose modern atomic theory of matter in 1803. The atoms were assumed as small solid sphere undergoing collisions, much like billiard balls. In 1897, J. J. Thomson discovered electrons. It radically changed the prevailing view of the atom. His work suggested the atom was not an "indivisible" particle as Dalton had suggested but an entity made up of smaller pieces. Even more revolutionary view came in 1911 when Rutherford suggested the possibility of a center or nucleus in the middle of an atom. This nucleus would have a dense core in the center. It would contain positively charged particles called protons. Surrounding the core would be mostly empty space wherein the negatively charged electrons would swirl around the nucleus. In 1932, James Chadwick discovered a third subatomic particle, which he named as neutron. Neutrons help stabilize the protons in the atom's nucleus and led to our current views of atom, a tiny solar system invisible to human eyes but firmly rooted in principles of quantum physics.

The word *quantum* means discrete or digital, as oppose to continuous. Max Planck in 1900 put forth the notion that the fundamental nature of radiation is not continuous. The electromagnetic radiation, according to his proposal, is emitted or absorbed in discrete quantities. He called these pockets of energy as quanta. Einstein building on Planck's ideas, in 1905, published that visible light energy also existed as "quanta" or discrete bundles of energy that behaved like particles, which he called photons. It was a revolutionary idea.

What nonsense! Energy, in discrete bundles! It was bizarre, certainly not an easy concept to accept. It shook the discipline of physics from its foundation. It divided physicists. It created quantum physicists. Rests were called classical physicists. It took decades of intense discussions and still the majority did not accept the proposition. The more heated the discussions got, the weirder became the consequences of adopting such ideas.

As the classical physicists learnt more about the nature at the smaller level, they found quantum or discrete or digital nature to be prevalent, in fact universally present. All the subatomic particles turned out to be quantum mechanical in nature. The atom, the invisible tiny solar system, itself could not have a stable structure without being a quantum mechanical entity.

The quantum physical view of the matter turned out to be nothing like the models earlier Newtonian physicists imagined. Instead, this view turned out

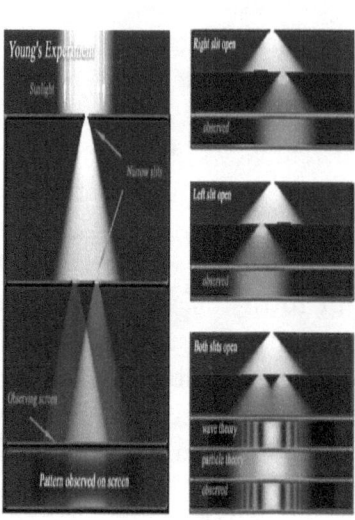

Young's double-slit experiment Is light wave or particle? The question was raised after the double-slit experiments were first carried out by Young in 1805 showed interference pattern indicating the wave like nature of light.

to be so weird that even quantum physicists were stunned to imagine its implications. It is hard to believe that such weirdness makes the foundation of us and our universe. But it is true!

## Weird Quantum Foundations of Our Universe

Since the atomic idea was finally accepted at the beginning of the twentieth century, atoms have proven central to the discovery and understanding of the laws of physics. Today remarkably sensitive techniques probe the properties of atoms, molecules, and light over enormous ranges: from submicroscopic to cosmic distances, in both familiar environments as well as in and the most exotic realms in the universe. Fascinating is the realm of quantum particles that make up the atom. A century of studies has revealed some of the strangest properties of these entities. Since all the visible universe is made from atoms, which in turn are made up of these quantum entities, the strangeness of this tiny universe pervades the nature of all there is, including us. So please pay attention.

### Quantum Entities, Wave Or Particle Or Just Information?

Our senses tell us that matter could either be a wave or a particle but not both. Quantum entities are both!

In our everyday experience, we know that waves are formed by motion within a medium. We are familiar with many different types of waves. My personal favorite is ocean waves. Ocean waves can be gentle and fun to watch or powerful enough to sink even a large ship. My son Monish, when he was little, would love to throw stones in water. He used to be mesmerized by the waves that would be generated when stones would hit the water. Sound waves roll outward from a source through the medium that could be water or air or may be a solid like a rope. A violin string waves back and forth, along its length, held in place at the two ends of the medium. A jerk on a loose rope sends a wave rolling along its length. Waves fascinate us.

We know a great deal about particles as well. Particles are solid objects with finite boundary like a ball, marbles, or bullets. Most of us like ball games, one form or the other. It could be baseball, football, cricket, or any one of the other

 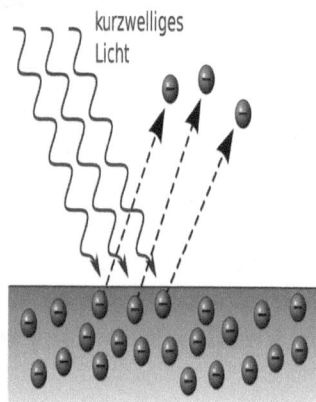

kurzwelliges
Licht

Einstein and photoelectric Effect. Einstein's explanation of the photoelectric effect rested on the assumption that light was quantized and behaved like a particle. These particles were photons. Finally when the mystery was solved, physicists had to contend with a strange compromise. Light is both wave and particle.

thousands that humans have invented and enjoyed since civilized. We are fascinated by the way the particles behave when struck by a force. We know waves and particles are different form of physical objects. Physical objects can be either wave or particles but cannot be both at the same time, especially if we trust our senses and logic.

All quantum entities defy this logic. All of them display both wave and particle like behavior. Is light wave or particle? The question was raised after the double-slit experiments were first carried out by Thomas Young in 1805 showed interference pattern indicating the wave like nature of light. Einstein's explanation of the photoelectric effect rested on the assumption that light was quantized and behaved like a particle. These particles were photons. Finally when the mystery was solved, physicists had to contend with a strange compromise. Light is both wave and particle. Quite weird, won't you say so? It almost feels like something quite obvious is missing, and physicists are not doing their job fully. It is not so. After incredible debates for many decades involving geniuses such as Einstein, the dual nature of this reality could not be simplified.

Not only light turned out to have this dual nature, so does every quantum particle. And not only that, so does every particle whether quantum or not. Everything in nature has wavelike properties as well as particlelike nature. If not observed, it is a wave. If observed, it becomes particle. Since larger objects are easy to spot and frequently observed, they appear to be particle most of the time. The smaller particles may spend more of their lives in wave state until an observer tries to see or know more about them. What about humans? Humans appear to be large particles, but the structures and processes that drive them to be alive have roots in quantum reality. Therefore, humans behave in a manner that resembles both large and small particles. We will further investigate this in details later in this book. First, let us examine the nature of the waves that these particles appear to emerge from.

It turns out that these waves are information waves, to be more precise, waves of potentialities. Later I would refer to these waves as the waves of guidance as they primarily guide the particle to occupy a preferred state of existence. These "waves" cannot be copied or even reckoned physically but can easily be calculated by a mathematical formula. This quality of something, having the

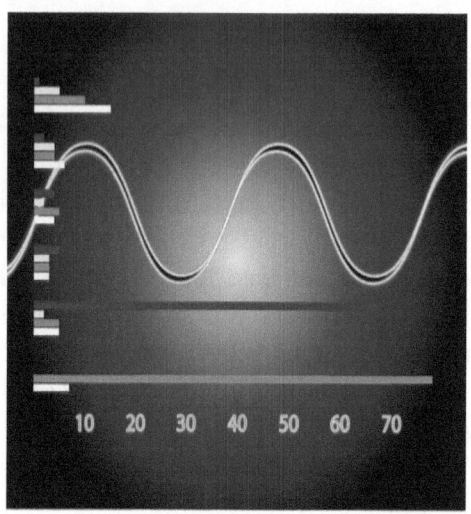

Quantum entities or information waves are nothing but information.

It turns out that these waves are information waves; to be more precise, waves of potentialities. One can also refer to these waves as the waves of guidance as they primarily guide the particle to occupy a preferred state of existence. These "waves" cannot be copied or even reckoned physically but can easily be calculated by a mathematical formula. This quality of something, having the appearance and effect of a wave but not the nature of a wave, is pervasive in quantum mechanics and is fundamental to all things in our universe.

appearance and effect of a wave but not the nature of a wave, is pervasive in quantum mechanics and is fundamental to all things in our universe. All quantum entities like electron, protons, neutrons, quarks, atoms, and even larger entities exhibit this behavior. The information waves appear to surround every physical entity in the universe, providing potential space-time options, related to the present and future states that are potentially available to the entity. Therefore, at any time, all entities seem to choose one among many options. As we will see later, that all form of simple or complex decision making, by the entity, can be reduced to posing of yes/no questions or binary choices such as 0 or 1. This is also the most basic form of quantum of action, where energy and information coincide. In a visible state, the quantum entity shows a definite binary value which could be either 0 or 1. However, in a wave state of unobserved particle, it stays in both states of 0 and 1 simultaneously. This is the most fundamental element of quantum information manifesting in the form of a computational entity, a quantum bit or qubit for short as will be described in details later. This nature of quantum particles forms the basis of a digital and computational universe. It shows how quantum entities, which are inexplicable, in physical terms turn out to be abstract, mathematical or informational in nature, as they form the digital and informational basis of us and our universe.

**Quantum Leaps?**

If you thought that dual nature of quantum particles was weird, wait till you hear what these entities do in real life. They leap around!

We all know about jumping from point A to point B. I am no sports personality, but even I have jumped a few times in my life. It means going from one point to another without touching anything in between. Most of us can do that. So if quantum entities are doing it, what is the big deal? The big deal is the transitional state. No matter how fast I could have jumped, I would have gone through countless transitional states before arriving at the destination. It is like if you took a video clip of my jump, you could watch me frame by frame. Each frame would represent a transitional point.

The transitional points, in fact, are very common. In our daily experience, we come across several such transitional states. In fact, when any property or a state of a physical entity moves from one state to the other, it goes through one or many intermediate states or middle states. For example, going from cold to hot, we go through a warm state. Or going from slow to fast one may go through many medium states. In general, a move from a state to another state involves a change in quantity until one reaches the other state. No matter how fast is the transition, intermediate states are involved.

Quantum entities move from one state to the other with no transition state in between. For example, one may observe electrons in a one-energy state in one instant and in other on the next. It seems to undergo no intermediate energy states. Similarly, the electron may spin one way in one instant and opposite the next. Once again, there appears to be no intermediate state. A pure stepwise process with no time or space present in any intermediate state.

It poses an intellectual challenge. How can it be possible that a physical object can leap from one state into another with no transitional state? Rigorous mathematical analysis of such scenarios is provided in what are known as Zeno's paradoxes. Through such paradoxes, Zeno proved that such change of state would not be possible. It in fact would require an infinite number of transitions if space-time were to be continuous.

Does it mean that space-time is discontinuous? Our modern view leads us to believe space-time as digital in nature, a grand network of discrete points with nothing in between. It appears that the digital nature of space, quantum information qubit-like behavior of quantum particles, form the foundation of computationally digital universe.

**Uncertainty, To Be or Not To Be?**

We embraced Newtonian classical physics because it made sense and was deterministic. We could measure location, speed, and momentum with precision and make predictions. It led to the industrial revolution. Indeterministic behavior was synonymous with ignorance or hidden variables as Einstein in his famous quote said, "God does not throw dice."

Careful observations of the behavior of quantum entities changed all that. Quantum world turned out to be inherently indeterministic. Properties describing quantum entities often came in groups. These are called complimentary properties. For example, location, speed, and momentum are complimentary properties. It turned out that these complimentary properties of the quantum entities could not be nailed down all at once. For example, when scientists measured the location and the speed precisely, the momentum could not be pinned down. Once the momentum was measured, one could determine the position. However, if one measured momentum again, one will find that the momentum had changed. There appeared to be inherent uncertainty involved in measuring these properties.

The obvious explanation was the act of observation itself. In determining a property, for example, position, a photon would have to bump into the quantum entity which could change its momentum. Many believed that better techniques and less intrusive instruments might solve the problem. This, however, did not turn out to be the case. Repeated experiments showed that the range of uncertainty of momentum increased in direct proportion to the accuracy of measuring location.

In 1925, Werner Heisenberg showed a mathematical incompatibility between these two properties. A mathematical relationship between the properties position and momentum was developed. It showed that the more precise your knowledge of the one, the less precise you can know the other.

The formula that estimated the uncertainty was certain. Heisenberg's mathematical derivation was verified experimentally, and it yielded far better correlations than the notion of needing better techniques or equipment in the laboratory. The uncertainty in the knowledge of two complementary properties was more than a laboratory phenomenon; it turned out to be a fundamental law of nature.

Our universe has turned out to be inherently uncertain. Which means the universe prohibits anyone to know precisely or exactly its inner workings. "No matter how bad it looks, it can always turn around quite unexpectedly!" may be the most fundamental law this quantum universe may live by. "Hope is alive and well and will always be so!" may be a positive way of expressing the grand uncertainty lying at the core of our digital and informational existence.

$$H(t)|\psi(t)\rangle = i\hbar\frac{\partial}{\partial t}|\psi(t)\rangle$$

*Erwin Schrodinger and his famous equation* What is quantum entanglement? Schrödinger discovered it through his newly formed quantum theory in 1928. Mathematically, when two quantum particles bump into one other, it was impossible to separate the two particles.

## Quantum Entanglement, Oneness?

The most amazing phenomenon at this size of quantum subatomic and atomic devices of our existence is entanglement. Erwin Schrödinger called this phenomenon the defining trait of quantum theory. Entanglement has a mystical touch to it. Its mysterious nature has challenged our sensibilities for several decades now.

What is quantum entanglement? Schrödinger discovered it through his newly formed quantum theory in 1928. Mathematically, when two quantum particles bumped into one other, it was impossible to separate the two particles. Once entangled, they behaved like one. A relationship of some sort was immediately established between the two particles. It is very much like us bumping into a stranger and instantly finding out that the stranger is very interesting and we develop a strong relationship immediately to the extent that we feel one with them.

Quantum particles appear to form that instant bond with each other every time by a mere act of bumping into one another. Unlike human relationships, the quantum particles form relationship based on pure love, trust, and truth. Once the bond is established, all the information about the particles, such as momentum or spin, becomes properties of both. No secrets at all. So if quantum state of one particle is changed, it will alter the quantum state of the other, irrespective of the physical distance between the two. Einstein found it frustrating and quite distasteful. He famously dubbed it *spukhafte Fernwirkungen*, "spooky action at a distance." That spooky connection between tiny particles seems to be everywhere, and its consequences affect the internal as well as the external world that we experience.

Today, it is possible to entangle any two particles; it could be photons of light, atoms, or even larger molecular structures. One can then physically separate them as far as one likes and observe effects of entanglement on each particle. This spooky, faster-than-speed-of-light association is one of the most fundamental aspects of not only just quantum science but all science. We know that quantum mechanics describes how atoms combine into molecules and so underpins chemistry. Since physical as well as chemical processes underpin all biological processes, could entanglement impact the emergent, macroscopic characteristics of life? The answer appears to be "yes". This relationship-building characteristics does not seem to be confined to just quantum or "microscopic" universe. New physics reveals that entanglement between particles may exist everywhere, all the time, and have recently found evidence that it affects the wider, "macroscopic" world that we live in. The entanglements probably exist not only outside of us but may even exist inside of us all the time. All our relationships, communication, and connectivity are ultimately rooted in this invisible link.

It has become a discovery of even far greater consequences, as scientists discover the mechanics involved in building relationships using these invisible links that seem to be everywhere. The understanding is coming from the advancements in the science of quantum information processing. As scientists find that this is not only a simple relationship-forming link but also is crucial to facilitate quantum communication! As the science of information processing has evolved rapidly for several decades, it has become clear that the entanglement leads to quantum information processing in a novel way. Binary information is stored in bits on binary processors such as silicon microchips, each of which can hold 0 or 1. A quantum bit uses quantum particles like photons or atoms as the quantum information stores. Each quantum particle can memorize infinitely long string of numbers. Entanglement means you can interact with these quantum bits (qubits for short) to process quantum information. Since our universe began with a quantum event, giving rise to a large number of quantum particles or events, all these particles may already be interconnected to form a massive quantum-computing system of incredible computing power. It perhaps takes this incredible computing power to create and run a complex universe such as ours.

Taking that bite into the fruit of scientific knowledge did enable us to know and understand our place among the neighbors that inhabited Garden of Eden. We have even become clear about various activities that sustain this Garden of Eden. It is becoming clear that the *silence* or *emptiness* is the ground on which this incredible garden was founded. Energetic entities ranging from quantum, subatomic, atomic, nanoscale, macroscale along with larger living entities including us human are the crops that grow on this fertile land. A key question, however, is, are these energetic entities truly separate entities that coexist in one universe? Or these computational, information entities are really parts of *one*, connected at a deeper level where they all exist within a giant quantum-computing network engaged in *one* grand, divinelike act of information processing?

# Chapter 5

# Computational Nature
## of Universe

## Dance of BITs, Bytes, and QBITs

Claude E. Shannon is widely recognized as the father of modern information theory. Shannon's pioneering work in the 1940s laid the foundations of information theory. No one, at that time, foresaw the far-reaching implications of his work. The information theory, today, not only forms the foundation of modern digital communication, but it also forms the foundation of the laws of nature.

Claude E. Shannon is widely recognized as the father of modern information theory. Shannon's pioneering work in the 1940s laid the foundations of information theory. No one, at that time, foresaw the far-reaching implications of his work. The information theory, today, not only forms the foundation of modern digital communication, but it also forms the foundation of the laws of nature. All the laws of physics turned out to be digital in nature and information at the root of our very existence. This has led to a notion that we are a part of a digital, computational universe which is processing quantum information at the core of its operation.

The laws of information eventually transform into laws of matter, energy, and forces. Therefore, any life or we humans for that matter are informational entities engaged in incessant computation at our core. This, of course, is quite a contrasting perspective to the one that is rooted in Newtonian materialism, where laws of matter and energy are the most basic laws of nature. Let us begin by a simple understanding of what is "information" and how does information processing relate to our physical universe.

We are living through times that are often described as the "Information Age." Our mind and psyche are inundated with information. All the television shows, books, and the Internet, there appear to be no end to the information that comes our way. Whether or not we know, at the most basic level, what exactly information is, we sure know how to receive, store, retrieve, display, print, code, decode, transfer, and manipulate it. We do it all day long. Simply put, we are all good at information processing. Since the information that most of us are familiar with is a manifestation of a system of computation called binary computation, I will begin our discussion with binary computing. Later I will introduce the concept of quantum computing, which is not so familiar to most but is perhaps the most powerful computational concept that humans have ever discovered.

## Binary Computing

### Binary Information Processing: An Abstract Reality

Before we tackle the issues related to information processing, let us discuss the very nature of information. We see information in a wide variety of complex

## Binary Math

- Binary math is based on powers of 2, as opposed to powers of 10 for decimal math.
— Given an octet (8 bits), when a bit in the octet is set (1) its value is . . .
— 128 = leftmost bit (most significant bit) = $2^7$
— 64 = next bit = $2^6$
— 32 = next bit = $2^5$
— 16 = next bit = $2^4$
— 8 = next bit = $2^3$
— 4 = next bit = $2^2$
— 2 = next bit = $2^1$
— 1 = rightmost bit (least significant bit) = $2^0$

**Binary Computation:** Binary information is present in our homes, offices, and cars. It is contained in hundreds of information processors that we use every day such as appliances, laptops, and desktop computers. The bank transactions, the television, music or video or telephone conversations all involve processing of binary digital information in the form of zeros and ones, represented by billions of billions of bits. Popular handheld devices such as iPhone, smart phones, PDAs, and MP3 players are all possible because of advances in binary computing.

forms. However, no matter how complex the information is, it is reducible to its simplest elemental form. Just as an atom is the smallest unit of matter, a bit or byte is the smallest unit of information, or binary information to be precise. A bit is a single numeric value either 1 or 0, which encodes a single unit of digital information. These two values are often interpreted as binary digits. In fact, the term "bit" is a contraction of binary digit. A byte is a sequence or collection of bits; usually eight bits equal one byte. At the most fundamental level, a bit is a logical variable that has two choices, almost like an abstract mathematical switch that occupies two positions on or off. This is opposed to a constant that has no variability.

Claude E. Shannon joined Bell Telephone Laboratories as a mathematician in 1941. One of his first assignments involved finding a solution to a rather practical problem. He was asked to maximize the efficiency of transmitting information through telephone lines. This would have allowed engineers to route many more calls through a single telephone line. The problem was of errors such as cross-connections. As the number of calls through a line increased, so did the possibility of some form of errors in communication such as cross-connection. Many of us are old enough to appreciate such problems in the early days of phone networks. The bottom line was that only so many calls could be routed through a line before the quality of experience of the caller or receiver would be sacrificed.

Shannon soon realized the similarities between Boolean algebra and telephone switching circuits. The switching of telephone circuits resembled the fundamental unit of information or Boolean two-value binary digits, or one bit of binary information represented by 1 and 0. The value 1 would mean "on," that is, the switch is closed and the power is on; and the value 0 means "off" or the switch is open and power is off.

Shannon's incredible insight was the way he looked at information. He saw information as something that helped one answer the questions. Just as in the popular game show named *Twenty Questions*. The show became popular in late 1940s as a weekly radio quiz program. The game starts with the selection of one player that is chosen to give answers. He chooses a subject which he keeps to himself. The objective is for others to find out the subject he selected. All of the other players ask questions. Each takes a turn. Every question can only be answered with a simple yes or no. Lying is not allowed. If a questioner guesses correctly, he wins. If twenty questions are exhausted without the correct guess, then the answerer has stumped the rest of the players, and he is declared the winner.

Although not guaranteed, the questioner often arrived at the correct answers. This would be very surprising to most. To an untrained mind, guessing and winning even once would be surprising, as the subject to be guessed could be anything. But the game illustrates how complicated information such as description of any arbitrary subject has a fair chance to be expressed in twenty or less yes or no answers or bits. Mathematically, it makes a lot of sense. If each question is structured to eliminate half the subjects, twenty questions will allow the questioner to distinguish between $2^{20}$ or 1,048,576 subjects. This provides a fair chance to make a correct guess. Shannon realized that all information could be expressed in the same manner or combinations of bits.

Tied to the basic binary choice may be a program or algorithm or a mind that selects the value that this variable can take. A bit could be connected to other bits and make a program or a control system or a processor or even a rational mind. Even a human free will may be tied to making that selection. Opposed to a choice, a binary switch (BIT) can have a value that can be probabilistically or randomly assigned or assumed.

Probability theory has played a meaningful role in developing modern human understanding. All disciplines of science including social behaviors and economic systems have benefited from this mode of understanding. It has been particularly noteworthy in physics. The classical application of probability theory gave us statistical thermodynamics. Statistical thermodynamics has been concerned with the macroscopic or collective behavior of large numbers of individual entities such as atoms or molecules. It attempts to correlate it to macroscopic phenomena that we commonly observe such as boiling or freezing or melting of water or the formation of raindrops. It was the connection between statistical thermodynamics and information that proved critical in establishing a link between the laws of physics and information.

One of the most influential contributions of Shannon's theory was the concept of information entropy. According to the second law of thermodynamics, developed in the nineteenth century, entropy represented the degree of randomness or chaos in any system. Entropy of an isolated system always increases. In the communication scenario, according to Shannon, it would mean that many sentences could be significantly shortened without losing their meaning. He defined information entropy to be equivalent to a shortage of the information content of a message. The use of entropy was clearly not new, but for the first time Shannon was using it in the framework of communication, especially as a tool to measure the quality of information. This groundbreaking work had far-reaching outcomes. It not only tied the well-developed science of thermodynamics to the laws of information but through this link, all the laws of physics, chemistry, and biology turned out to be rooted in the broader laws of information. It did not happen overnight. It took decades of heated debates in the scientific community. This is how it started.

The failures of caloric theory, and greater acceptance of the heat content of the system to be related to its internal microscopic matter in motion, prompted many question about the relationship between classical mechanics and thermodynamics. One of the most notable one was, does thermodynamics apply to everything including microscopic systems? With the incredible success of the laws of thermodynamics, there was no reason to doubt that it would apply at the microscopic level too. R. Clausius in 1850 and then Boltzmann in 1877 introduced entropy in statistical mechanics, which considered molecular structure of matter in its formulation. The concept of entropy was already well established in thermodynamics as a measure of chaos, disorder, or randomness in systems. Boltzmann developed the second law of thermodynamics in the framework of statistical thermodynamics using statistics and probabilities. He was even able to prove that the second law must be true within statistical certainty. The absolute certainty was the hallmark of classical physics then. Any uncertainty was assumed to be because of some degree of ignorance or hidden variables. The probabilistic nature of Boltzmann's work made it appear that the absolute certainty of the laws of thermodynamics was undermined. This perception of lingering ignorance was the demon that followed Boltzmann throughout the rest of his life.

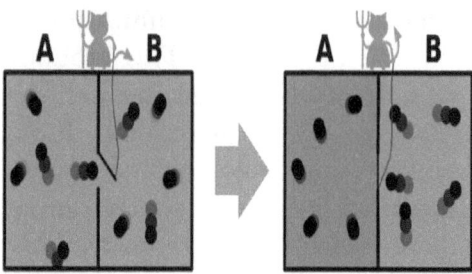

In 1866, J. C. Maxwell's neat-fingered being became famous as "Maxwell's demon." It could lead to a potential violation of the second law of thermodynamics. If demon was intelligent, it would simply exploit the random motion of the molecules, and allow them to sort themselves, by simply letting the molecules through the gate at the right time. In other words, the intelligent tiny being would allow the molecules with higher energy from cold chamber to go to hotter chamber making cold chamber even colder and hot chamber even hotter.

In 1866, J. C. Maxwell expressed this demon in the form of a thought experiment. His famous "neat-fingered being" could lead to a potential violation of the second law of thermodynamics. The demon could make a hot system hotter and a cold system colder without doing any work. The being became famous as "Maxwell's demon" and was called for it to be expelled to preserve the integrity of one of the most successful laws of classical thermodynamics. According to Maxwell, a little intelligent being (the demon) could be deputed at the interface or a gate that connects two chambers, one filled with hot gas and the other filled with cold gas. If one starts out with hot and cold atoms mixed evenly in two chambers, then the demon's goal would be to end up with all hot atoms on one side and cold on the other side. To violate the second law, demon would have to avoid performing any work to carry out this task. In other words, if demon was intelligent, it would simply exploit the random motion of the molecules, and allow them to sort themselves, by simply letting the molecules through the gate at the right time. In other words, the intelligent tiny being would allow the molecules with higher energy from cold chamber to go to hotter chamber making cold chamber even colder and hot chamber even hotter. A neat trick! If true, it could form the foundation of a perpetual motion machine. None of us would have to worry about energy then.

After decades of lively debate, the Maxwell's demon was dispelled but not before it reduced the laws of thermodynamics to a special case of essentially much broader laws of information. C. H. Bennett explained Maxwell's demon by noting that to decide what side of the gate a molecule must be on, the demon must store information about the state of the molecule. Eventually, the demon will run out of information storage space and must begin to erase the information, and by Landauer's principle this will increase the entropy of a system. Landauer principle stated that any logically irreversible manipulation of information (for example, erasure of a bit) must be accompanied by a matching entropy increase in an isolated system. If no information is erased, a state may in principle be achieved that is thermodynamically reversible and requires no release of heat.

Shannon's information entropy and entropy of thermodynamics were same or at least intimately related. His information theory radically changed the way scientists looked at the universe. It created a great revolution in physics of twentieth century, *an information revolution.*

# Binary Information Processing: A Physical Reality

The logical variable bit, a string of 0 and 1, in the true sense means something numeric, mathematical, or abstract. It may be surprising to many that an abstract concept such as a bit can actually be represented or simulated by one or more physical devices. In fact, the physical devices that represent bit are common and everywhere. Simple examples are on/off switches or a key lock combination that we use in our daily lives. Some more not-so-obvious examples include tiny charges on a transistor, micron-sized patches of magnetic material, and microscopic burn marks on a CD or DVD. These all represent physical bits. The operations of devices that contain these bits result in computation involving one or more binary variables. In fact, the storage and processing of information always requires some kind of physical means, such as the physical position of a switch or the electrical charge on a capacitor.

Binary information is present in our homes, offices, and cars. It is contained in hundreds of information processors that we use every day such as appliances, laptops, and desktop computers. The bank transactions, the television, music or video or telephone conversations all involve processing of binary digital information in the form of zeros and ones, represented by billions of billions of bits. Popular handheld devices such as iPhone, smart phones, PDAs, and MP3 players are all possible because of advances in binary computing.

A new report released by the University of California, published recently in *The New York Times* Bits Blog, claims that American households collectively consumed 3.6 zettabytes (one sextillion bytes) of information in 2008. Individually, the study claims the average American consumes 34 gigabytes of content and one hundred thousand words of information in a single day from PCs, TVs, radio, video games, and text messages. It seems like our appetite to crave and digest binary information is on the rise. Even us humans often make decisions in the form of 0 and 1 or one may say go or no go. For example, you may turn a television on or off. By turning it on, it starts a series of processes that manipulate information with results of you watching the television programs. However, all conventional physical bits share one defining feature: a bit is in one state or the other. In other words, it is either zero or one but never both. This is what differentiates binary computing from quantum computing.

## Quantum Computing

When a bit is both zero and one at the same time, it is no longer a bit; it becomes quantum bit or qubit for short, the smallest element of quantum information. Quantum information is different. It is stored, not in bits, but in qubits. How can you have a physical switch both in on and off positions at the same time? This may challenge your sensibilities. When I first shared this with my wife, Seema, during a casual evening walk, she looked at me as if I was inebriated. "What kind of science fiction is this?" she exclaimed. Except that it is not science fiction; it is a science fact and perhaps the most fundamental reality of our existence.

An ordinary, large transistor cannot be both on and off. But if it is small enough, the rules of quantum mechanics may take precedence, making such an oddity not only possible but typical. Thus, a single atom can be in what is known as "superposition" of two different states. The superposition is one of the most amazing and almost mystical states of existence. There is no physical equivalent of this state. "There is something as well as there is nothing." From a physical point of view, it could be called unmanifested state of being.

Just as binary bit can be simulated by physical devices, quantum bit also has physical devices that simulate the nature of this variable. In fact, all quantum entities represent this variable. For example, an atom's outermost electron may spin up or down, or it can be in a superposition of both up and down states. An essential and nonintuitive feature of this superposition is that in this state, the spin axis is not someplace between up and down. If we observe the spin, it will always be seen to be either up or down, never someplace in between. However, this is not because the spin is indeed up or down before being measured; it is truly in both states and does not assume in one or the other state until it is measured.

This may sound illogical. However, it is a central principle of quantum mechanics. The most basic description of nature, at its most fundamental level, implies that systems can be in two or more configurations at once.

It is only when one observes the system that it will collapse randomly into one of its possible configurations.

Computers are physical objects and
computations are physical processes.
What computers can and cannot do are
determined by the laws of physics alone
and not by pure mathematics.

—Deutsch

Since our universe is ultimately made up of these quantum entities, the *whole* universe may share this common invisible superpositional state of being. I will describe this informational entity, later in this book, as *whole* which represents all that is visible and invisible. One may suggest that all devices of this universe are ultimately naturally connected as one *whole*. Or there is only whole with such devices as its parts.

Binary bits network to form a binary computer or a network of computing systems. Essentially the process is to form local binary computing system, followed by interconnecting these through binary networks, much like the Internet today. We are quite familiar with this mode of building binary architectures. Quantum computing throws this concept upside down. It means that all that needs to be networked is already networked, and localized computing evolves subsequently. It means that all quantum information is already networked and has been so from the beginning. All qubits share this grand state of wireless networking.

At the most fundamental level, a qubit is a variable just like a bit. This is opposed to a constant which has no variability. Tied to the basic three choices may be a program or quantum algorithm or a mind or a quantum computer that down select the value that quantum variable can take. A qubit can be connected to other qubits or bits and make a program, a controller, a processor or a quantum computer, or even a life form. Or just like a bit, a human free will may be tied to making that selection.

What makes quantum information so special? Superposition allows qubits to do things that ordinary bits cannot. To see why this is so, consider a register of three classical bits. More precisely, that three-qubit register can be in a coherent superposition of all eight numbers. The above superposition is "coherent" because its weights have definite phase relationships between them. This allows interference to occur, much like the interference of any wavelike phenomena. This allows exponential scaling of quantum information, a feature that is not possible with binary information.

The exponential scaling of quantum information is related to one of the weirdest aspects of quantum theory—"entanglement," as described earlier. This feature is so strange that in 1935, when quantum mechanics was in its infancy, Einstein and two colleagues wrote a paper pointing

> Every finitely realizable physical system can
> be perfectly simulated by a universal
> (quantum) Turing machine operating
> by finite means.
>
> —Feynman-Deutsch Principle

out just how strange it is. It became well-known as the EPR paradox (or Einstein-Podolsky-Rosen paradox) after the initials of the names of the three authors of the paper. They argued that nothing so strange could be true, and instead there must be something wrong with quantum theory. In fact, this was a rare instance where Einstein and his colleagues were wrong; the nature indeed proved itself out to be that weird. This weirdness, however, enabled quantum computing and provided firm footing to the science of quantum mechanics. Just how does that happen? Let us further examine with a simple example.

Consider just two qubits, each of which can be in one of two states. We can think of these as two atoms, labeled A and B, whose spins could be either up or down along some chosen direction.

The first of these states is said to be "separable" because it can be written as the product of the state of one atom and the state of the other. If we measure the spin of one of the atoms, it will be either up or down.

For example, we might measure atom A to be up and find B to be down. Or we might find that A is up and B is also up. The result of each measurement will be random, and the result for one atom will not depend on the result for the other. This is a characteristic of separable states.

In contrast, the second state cannot be written as the product of the state of one atom and the state of the other. This is an "entangled" state. In this state, each atom is in a quantum superposition of being up and being down. If we measure the state of atom A (or of B), it may either be up or down. The result will be random. But after measurements, A was in up state, then for the state B is bound with certainty to be measured as up. Conversely, if A is down, B will definitely be found to be down. Therefore, each atom can assume a value randomly upon observation, but their fates are linked. This faster-than-light link or communication is what we mean by entanglement. The communication between the two states is called nonlocal, which means that each of the two atoms could be at the two ends of the galaxy, and still this communication will take place in an instant. As soon as one is observed, and assumes a state of on or off, the other will instantly know and assume the other state.

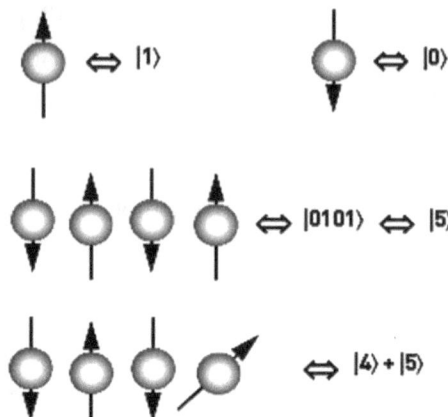

**qubits can be in a superposition of all the classically allowed states**

Quantum BITs or QBITs

What makes quantum information so special? Superposition allows QBITs to do things that ordinary BITs cannot. It allows interference to occur, much like the interference of any wavelike phenomena. This allows exponential scaling of quantum information, a feature that is not possible with binary information. It allows for a quantum-computational system of incredible power and capacity. For a modest 300-QBIT register, the number of states in the superposition can be 2 to the power of 300, a number that is enormously larger than the number of atoms in the entire universe.

The computing capacity gained by allowing the qubit register to be in a quantum superposition is enormous when the number of individual qubits becomes large. The number of states allowed in the superposition grows exponentially with the number of qubits in the register. For the three-qubit register, the number of possible states in the superposition is 8 or 2 to the power of 3. The three-bit register stores the binary number 101, which is 5 in the usual decimal system that most of us are quite used to. A three-bit classical register can store any one of eight different numbers from zero=000 to seven=111. In contrast, a quantum three-qubit register can store a superposition of all the eight different numbers at the same time.

Computations performed on a string of qubits are thus the equivalent of computations performed on every possible value of a string of ordinary bits. Since eight classical qubits can represent 256 different values, a single eight-qubit calculation is the equivalent of 256 classical calculations or 2 to the power of 8. By simply adding another qubit, one can double the number of simultaneous classical computations the quantum computer is able to perform. Thus, superposition of quantum-computing elements allows a quantum computer to perform multiple calculations simultaneously in parallel. The classical computers cannot duplicate this feat. In fact, classical computers often fall short to efficiently simulate a large quantum computation.

For an N-qubit register, the number of states in superposition is 2 to the power of N. Thus, while a classical N-bit register can represent a single N-bit number, a quantum N-qubit register can be in a superposition of all 2 to the power of N numbers. For a modest 300-qubit register, the number of states in the superposition can be 2 to the power of 300, a number that is enormously larger than the number of atoms in the entire universe. It shows how powerful quantum computers can be. Seth Lloyd, professor of quantum mechanical engineering at MIT, computed the limit on the amount of information that the universe can register and the number of elementary operations that it can have performed over its history. His calculations reveal that our universe could have performed no more than 10 to the power of 120 operations on 10 to the power of 90 bits. These numbers are of course mind-boggling, kind of shows our place in the universal scheme of things!

Three classical bits register of binary values 101: The three-bit register stores the binary number 101, which is 5 in the usual decimal system that most of us are quite used to. A three-bit classical register can store any one of eight different numbers from zero=000 to seven=111. In contrast, a quantum three-qubit register can store a superposition of all the eight different numbers at the same time.

Besides this ability to store exponentially many quantum states, the linear nature of quantum mechanics means that these states can all be manipulated at the same time—that is, massive "quantum parallelism" is possible. This is another key to realize the power of quantum computation. It allows a quantum processor to perform an exponentially large number of calculations all at the same time, since the quantum register contains that large number of different classical registers. This can make a quantum computer mind-bogglingly faster than even the fastest imaginable classical computer.

A problem that could take more than a human lifetime to solve on the best classical computer of today might be solved in minutes or hours on a quantum computer. The best example of this today is the problem of factoring a large number into its primes, a computational problem that is extraordinarily time-consuming on a classical computer. In fact, the factorization of large numbers is so hard that it forms the basis of most data encryption standards today. In 1994, Peter Shor showed that a quantum computer would be capable of doing this task exponentially faster than any known classical algorithm, that is, the solution time using the classical algorithm grows exponentially with the number of qubits required to represent the number, while the corresponding time for a quantum solution grows much more slowly. Consequently, quantum computers could, in the hand of a computer hacker, compromise the security of many forms of encryption in use today.

Recently, quantum mechanical engineering professor Seth Lloyd, who along with Avinatan Hassidim, a postdoc in the Research Lab of Electronics, and the University of Bristol's Aram Harrow came up with this new quantum-computing algorithm that would significantly reduce the steps it takes to solve equations that have a large number of variables. The computation time required by their quantum algorithm would be proportional to the logarithm of the number of variables. In contrast, computation time required by a classical binary computer is proportional to the number of variables. It means that for a calculation involving a trillion variable, "a supercomputer is going to take trillions of steps, and this algorithm will take a few hundred," says Seth Lloyd.

Quantum computing machines enable new
algorithms that cannot be realised in a classical
world. The algorithms can be powerful physical
simulators. The physics determines
the algorithm. The hardware matters.

—G. J. Milburn

When I described this to my older son Vinay, who was taking AP high school math classes, he was visibly disturbed. He exclaimed, "It sounds interesting, Papa. I am trying to solve these equations with just a few variables, and that is hard enough. Why would someone worry about trillions of variables?" The fact is, when modeling any natural systems, the variables add up. Even trillions of variables represent a small number. A series of chemical reactions, or modeling weather patterns, all require incredibly large number of variables. It is truly mind-boggling how our physical universe computes orders of magnitude larger number of variables every instant, all the time.

From a quantum mechanical perspective, we seek the smallest element of action or information or one may say an event. We can picture our visible world or everything it contains as sequences of interconnected events or actions, each of which can be subdivided into smaller events. Let us say I have thrown a party. All my guests are eating drinking, singing; and some are dancing, having a fabulous time. Thus one way to describe my party is as a collection of interactions of a roomful of people, talking, eating and drinking, and so on. A particular conversation may be broken down into a series of utterances, each of which is an event. Each utterance can further be divided into smaller and smaller events until one reaches the smallest unit of quantum information, a qubit.

Qubit or a tiny "quantum of action" is deeply rooted in local as well as nonlocal action. This action or event is so tiny and rapid that even the simplest event in human space comprises a vast number of these elementary events. These events, though distinct, weld into a larger seamless whole, which could be my guests or me or the party that I have thrown. My party itself is part of my life and others who joined in the event, and each of our lives is part of the history of the human race and so on. Therefore, every current event and all past events are connected in local and nonlocal way through incredible numbers of elementary quantum informational events.

The quantum transformations take place in superpositional or unobserved state of our existence and manipulate the probabilities of all events in observable space. At this tiny level, each particle knows what every other particle is doing at all times. Most scientists believe that our universe began with a quantum event, big bang, eventually giving rise to a large number of quantum particles or events. Going back to the analogy of the universe

The results of all finitely describable physical measurement systems can be perfectly simulated by a universal quantum computer operating by finite means, producing finite measurement records.

—Feynman-Deutsch Principle
(Measurement Formulation)

as a giant quantum-computing system, the big bang would be the onset of a new quantum-computation platform or the start of an incredible dance of bits, bytes, and qubits.

At the onset, incredible number of qubits or elementary units of quantum information were created. All networked together in a grand superpositional state as the dance of these qubits and bits began. Initial dance moves were simple patterns. As time passed by, the movements became more complex, intricate, and engaging. The qubits combined to form complex informational entities. As the dance continued, more complex patterns emerged. As these informational entities interacted, and further connected, even more complex structures emerge. Life and humans are just one of these structures. Therefore, all these informational entities interconnected or entangled to form a massive parallel quantum-computing network of incredible power. This logic defying amazing computerlike superpositional state of existence is the true state of our universe. It is energyless, timeless, invisible, and *one* at its core. A state of silence or emptiness and yet filled with incredible nonlocal computations. All the entities including us in our visible universe arose from this invisible quantum-computing state of existence, leaving us to ponder: "Is there a grand quantum computing nonlocal mind hidden in the collective emergence of entangled information wavelike states of all quantum entities?"

Before we examine the nature or even an existence of a nonlocal cosmic mind, let us examine the possibility of a local mind which is allowing us to read and ponder the possibility of a cosmic mind or digital divine. Laws of information seem to govern our universe; however, what it is that converts randomness into information? Is there a local mind responsible to turn random into information?

# Nature of
# Local Mind

# Chapter 6

# Rational Mind

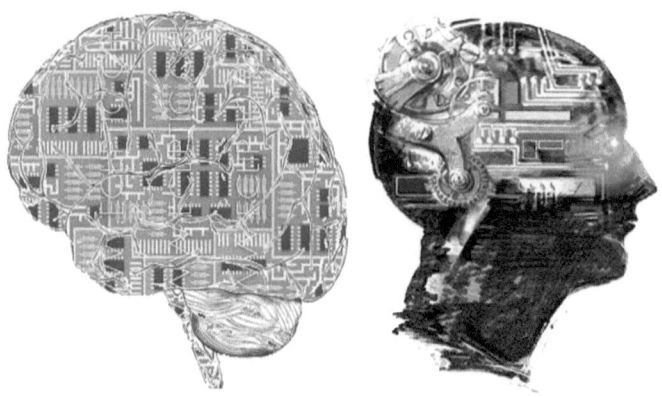

Rational Mind or Binary Computer

"Can the operations of the brain be simulated on a digital computer? . . . The answer seems to me . . . demonstrably 'yes' . . . That is, naturally interpreted, the question means: Is there some description of the brain such that under that description you could do a computational simulation of the operations of the brain. But given Church's thesis that anything that can be given a precise enough characterization as a set of steps can be simulated on a digital computer, it follows trivially that the question has an affirmative answer" (Searle, 1992).

## Triumph of Reason, Rational Mind

In an uncertain universe, randomness rules, and rational mind aims to bring the element of certainty through reason. In fact, our modern life cannot be defined without reason or rationality. Rationality distinguishes us humans from many other living species as the ability to reason is either rudimentary or outright lacking in other living species. It is, in fact, our defining trait. Among all talents, the ability, which most marks us as human, is our capacity to think, deliberate on, analyze and classify facts, develop ideas, and decide on a course of action. These are also the skills of a developed, rational mind. All our education is rooted in reason. Our justice system is deeply rooted in the notion of reasoned right and wrong. Reason is the cornerstone of our criminal justice system. Reasonable doubt is the edge of reason. Beyond the edge, on one side, there is reason. There is no doubt. There is certainty. Guilty as charged! On the other side, there is uncertainty, doubts, unknowns, or randomness. The question still remains, what is the nature of reason? Who decides the reason?

## The Egde of Reason, A Case of Reasonable Doubt

In 1991, two young women were reported missing after visiting the abandoned Chain of Rocks Bridge in St. Louis Missouri, a popular hangout for local teens. The young boy, who was with the young women that day, told the police an interesting tale. He told cops that the young women were pushed from the bridge by an unknown assailant. He was also ordered to jump but survived the nearly eighty-foot fall into strong currents. At the time of the interview, he had no injuries, and his hair was dry. The police were naturally skeptical of his account, and within hours, he confessed to killing the young women.

Yet this man, who is white, has never spent a day in jail. Instead, the police arrested four local youths who were also on the bridge that night. Three of the young men, all African American, received the death sentence. The fourth young man, who is white, received a thirty-year sentence and will be eligible for parole soon.

Reggie Clemons is one of the youths that received the death sentence, even though prosecutors conceded that Reggie neither pushed the women

113

nor planned their deaths. The prosecutor theorized that Reggie was an "accomplice" even though there is no physical evidence linking Reggie to the crime for which he received the death penalty: no fingerprints, no DNA, any hair or fiber samples.

Many of Reggie's claims have never been heard in a court of law because of procedural rules that have barred presenting important evidence. After reviewing the evidence, two federal judges voted to overturn his death sentence and found that Reggie was denied a fair trial.

It begs the question, "Is our justice system governed by perception of communally agreed-upon reasons, or is there an absolute reason?" Even though the decision and, therefore, the reason may not be absolute, it is crucial for our modern existence. In fact, every facet of modern human society relies on reason to preserve its existence and advance its cause. It would not be an understatement to credit today's human progress as a triumph of reason. Moreover, a great deal of the responsibility rests on one evolutionary growth—our rational mind. How our rational mind has evolved over the last several billion years with such skills to bring reason out of randomness or chaos is an incredulous story.

## From Random to Reason, Information and Mind

*Random* is a great word. It is one of my favorite words. A Google search revealed 330,000,000 hits. What does it mean? No context. Out of the blue. Odd. The dictionary.com defines it as proceeding, made, or occurring without definite aim, reason, or pattern. What about the word *reason*? We seem to like it almost as much or more than the word *random*. How can I say so? Well, it showed 459,000,000 hits in similar Google search. So by some "reason," one might say, it is more popular than the word *random*. The same site defines reason in terms of mental powers concerned with forming conclusions, judgments, or inferences. What about information? I know if I Goggled *information*, I would get 162,000,000 hits. Not bad! It may not be as popular as the word *reason* or *random*, through some reasoning standards, but we know it is critical for our modern existence. There is no denying that we live in the information age and process information all day long. Even all the cosmic laws appear to be reducible to laws of information. Modern science of information is leading us to believe that these three words are intimately related. What connects these words? The answer is one of the most influential evolutionary growths responsible for origin and development of all life, the mind or more specifically our local mind. Without it, our existence would not be possible.

It is a trait of living organisms to gain information, interpret it, and pass it on, often using it and refining it along the way. What distinguishes living from nonliving is the way the living entities are able to process information. This information can be available in various forms. For example, it can be genetic information passed on from the parent to the offspring, sensory information transferred by a sense organ to the brain, linguistic information, or numerical data entered in a computer to facilitate computation.

It is easiest to quantify information using a framework of communication. When one person sends a message, the measure of the information contained in the message may be measured by the increase in the knowledge of the second person on receiving the message from the first. The larger the number of possibilities contained in a message to narrow its content, the more is the misgiving removed on its receipt, and, therefore, the more is the information contained in it. The simplest message would be just a yes or no, distinguishing between only two possibilities, a bit or binary information as described earlier. As described in the last chapter, Claude Shannon measured the information, contained in a message, in the form of information entropy; which is simply the number of possibilities contained in the message to suggest a meaning. A repetitive message would waste resources repeating what is already suggested. So information contained in a message is increased by removing correlations among its parts; as messages become more efficient, they appear more and more random.

Shannon's work tied the entropy and information closely to each other. Entropy, the chaos or randomness in the system, is in fact a measure of information. Information lives in randomness, but it is not random. What distinguishes information from randomness is the sense of purpose, a reason the message has a meaningful interpretation for the receiver. Of course, that requires a common language, which both the sender and the receiver understand. A mind or perhaps a rational mind plays a key role in converting randomness into information.

115

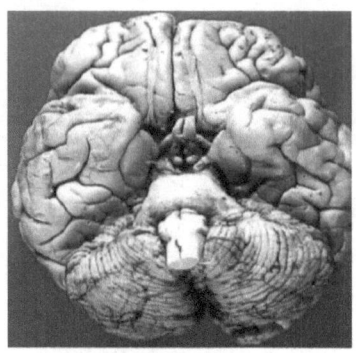

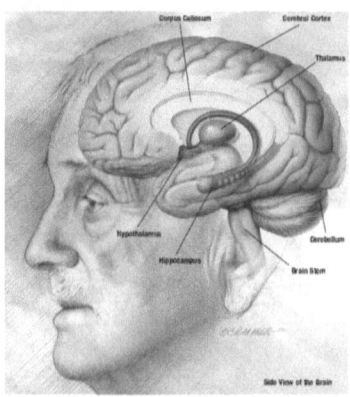

While a new chip in personal computers is introduced almost every year, it took the cosmic mind almost a billion years to build the neocortex. With the addition, the mammalian brain became the human brain. It added massive amount of gray matter, which enveloped most of the earlier brain. In the end, it amounts to about 85 percent of the human brain mass. It brought in a lot of new capabilities and features. It allowed humans to cope with parenthood and promoted social interactions. In fact, we have like modern human beings because of our neocortex.

According to R. Feynman, "Information is not simply a physical property of a message: it is a property of the message and your knowledge about it." At the very least, the information needs a receiver and a processor of the information to decipher its meaning. In other words, information needs an information processor to complete its circle of existence. In other words, a mind as an information processor can complete the circle of existence for information by catalyzing conversion of randomness to information. Of course, it needs to supply that reason, and the quality of reason ultimately determines the quality of information and thus communication. In general, all form of information is eventually tied to a mind, whether it is cosmic mind, emotional mind, or rational mind. In fact, information is at the root of reason, and, therefore, it is at the root of mind itself.

The simplest form of binary reason is contained in the nature of bit itself. The simplest form of quantum reasoning is contained in the nature of qubit itself. All information can be simplified to strings of interconnected quantum qubit or binary bits. A rational processor of information needs at least some information to be reduced to binary format, which could be tied with a reason. For humans, that binary information processor is our rational mind. It lives in our head. It is intimately connected to our emotional mind, through intricate informational network, that extends throughout our bodies, so every cell of our body can experience the impact of this rational processor. Reason that governs this processor is often contrasted with authority, intuition, emotion, mysticism, superstition, and faith. The meaning of the word *reason* overlaps to a large extent with *rationality*, which is normally *rational* rather than *reasoned* or *reasonable*. Reason and therefore information not only forms the basis of all laws of matter and energy, but it also seems to be at the root of our rational mind. Shannon's simple but elegant model of treating information, forever, changed the way we looked at nature.

## Rational Mind or A Binary Computer

Our human, rational mind organizes information from randomness. How do we know this? A clue comes from working chemicals that induce experience of deep sleep such as anesthetics. Most of us believe that anesthetics work by putting us to sleep through blocking receptors. However, most anesthetics do not work this way. They, in fact, introduce a noise or randomness or nonreason into the brain to confuse the neurons to prevent us from making "decisions." Our neurons must make scores of binary decisions to sense even a simple object. When we look at a room, we see an image of the room on the screen of our consciousness. "Almost 99% of what we see in a room is not what comes in through the eyes—it is what we infer about that room," says Henry Markram, founder of the Brain Mind Institute in Switzerland. Our rational mind creates, builds, informational entities that help us form a sense of reality, allowing us to navigate in the physical universe. How do the neural connections inside our brain lead us to form such a reality? To understand, we would need to understand how our brain has evolved.

The human brain has evolved in three main stages. First, the reptilian brain, the most primitive part and its innermost core. As the name suggests, it has remarkable likenesses to its structure and functions of brain present in reptiles. This reptilian brain, in our human body, controls basic body functions needed for carrying life such as breathing. Our behaviors about our survival, such as seeking food, shelter, or sexual activities are also controlled through this part of the brain. The response of this brain is rapid, instinctive, and automatic.

As mammals evolved from reptiles, enormous evolutionary changes took place in the body and in the mind. The reptilian brain changed and added what we call as the mammalian brain. It is like adding additional microprocessors in personal computers. You may recall, a few years back, each of the personal computers used to have just one microprocessor. Nowadays, it is common to see dual-core processor or quad-core processors in personal computers. The evolutionary transitions must have been similar to adding extra computing capacity leading to extra skills for mammals. The mammalian brain changed to control body functions such as blood pressure, the fluid balance, and body temperature. It used autonomic nervous system, hypothalamus, hippocampus, and amygdala to create and store experience-based memories and feelings about events. The feelings such as attachment, anger, and fear emerged, associated with behavioral patterns of care, fight, or flight.

Just as personal computers continue to improve in their computation power by adding more processors, mammalian brain also continued to add computing skills through the evolutionary process. The result was a third part of the brain called the neocortex. While a new chip in personal computers is introduced almost every year, it took the cosmic mind several hundred million years to build the neocortex. With this addition, the mammalian brain became the human brain. It added massive amount of gray matter, which enveloped most of the earlier brain. In the end, it amounts to about 85 percent of the human brain mass. It brought in a lot of new capabilities and features. It allowed humans to develop enhanced skills to cope with parenthood and promoted social interactions. In fact, we behave like modern human beings because of our neocortex.

The neocortex consists of two hemispheres. Each of the hemispheres is covered with an outer layer. Thick bundles of nerve fibers connect these hemispheres, facilitating communication between the two. The left part controls the right side of the body, while the right part controls the left side of the body. The two hemispheres are intricately interconnected and constantly in communication with each other.

While the left hemisphere controls verbal skills, the right hemisphere controls nonverbal skills. The left hemisphere deals with language involving choice, meaning, and rules of putting the words together. The right hemisphere seemingly controls the emotional content of speech. Left hemisphere is logical, systematic, and is concerned with matters as they appear. Right hemisphere, on the other hand, communicates using images. It has developed spatial, intuitive, and imaginative abilities. It is concerned with our emotions and feelings. All these remarkable abilities and skills are brought together as a comprehensive whole, the human information processor, our mind. Its ultimate efficiency and functionality depends on the way its parts contribute and cooperate with each other to form a whole.

This new brain is an expert information processor, comprising of a sophisticated network of neural connections. Its repertoire of functions includes the processing as well as memorizing of images, developing language, carrying out mental processes connected with memory and memorizing and facilitating skills for objective and logical thinking and evaluation. It also creates and manages a wide range of emotions or feelings involving love, care, and affection.

Just like the sensors and receptors found on the cell membrane of each living cell, the task of mind is also to sense and keep the whole organism aware of the environment and make suitable decisions. Neural connectivity is at the

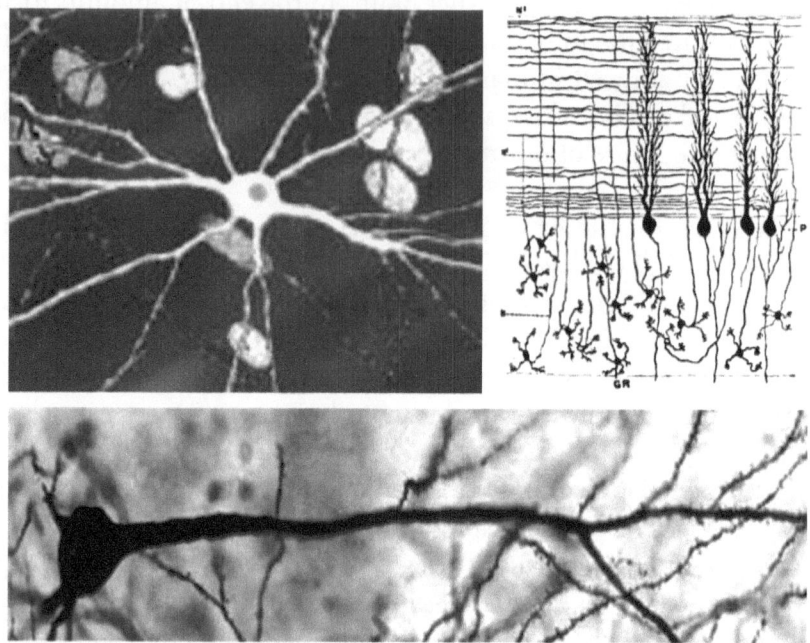

Human neural system is digital in nature. Nerve cells have evolved to develop a mechanism that allows them to transfer information. The input signals may come from various sources, for example, other nerve cells or from sensors implanted on cell walls as well as the dendrites. Each nerve cell may be receiving one or many inputs at a time. Inputs lead the cell to create an electrical pulse.

root of information processing and decision making. Neural connections evolve as we go through life. Neural connectivity is like developing friendship in humans. Just like humans, who stay together, talk, or do activities together tend to become friends, neurons that fire together also end up connecting. In fact, this neural chatter when repeated leads to such strong bonding that they turn into conceptual node or memory. Such association can be recalled by accessing just about any of the engaging neurons by simple intentions of recalling memory connected to the engaging neuron.

Human neural system is digital in nature. Nerve cells have evolved to develop a mechanism that allows them to transfer and process information. The input signals may come from various sources, for example, other nerve cells or directly from sensors implanted on cell walls as well as the dendrites. Each nerve cell may be receiving one or many inputs at a time. Inputs lead the cell to create an electrical pulse. It travels, away from the nucleus, toward the end of axon and eventually to the ends of the terminal branches. There, it is broadcasted to other nerve cells.

On receipt, some pulses tend to excite the receiving nerve cell into creating an electrical pulse and sending it down the axon. In some cases at other locations, the nerve cells may inhibit the generation of a pulse. It would stop the signal from propagating from that node. In other words, the nerve cell acts like a decision mechanism, which creates pulses in response to a form of signal summation, very much like a multiterm Boolean function in modern computers. The signal is then delivered to other nerve cells for further computation.

The axon signal involves an electrical voltage pulse, remarkably similar to our modern digital signal communication. Our modern communication systems handle high volume data transmission, often by multiplexing single paths. A single channel, for example, a fiber optic line, is set up on a signal conductor, and volumes of signals are processed by time-sharing this line. The signals are switched so rapidly that it provides appearance to the end user that a separate, dedicated line was assigned to each signal. Electronic signals, in our modern communication network, travel at the speed of light or 300,000,000 meters per second. However, the speed of the electrical pulse that carries the signal in an axon is roughly about 100 meters per second, an order of magnitude slower. Switching speeds on electronic equipment is in the order of nanoseconds or 1/1,000,000,000th of a second. Axon signal switching speeds, on the other hand, are in milliseconds or 1/1000th of a second, about a million times slower.

Clearly our nervous system ranks extremely slow in comparison to our modern communication systems. Even though it is slow, but it is far superior to any current modern information processor. Its ingenious design overcomes many of its limitations. For example, consider the human eye. The eye is composed of thousands of photoreceptors cells. A separate nerve axon connects each photoreceptor to the brain. There are thousands of nerve cells working in parallel. It can by itself, in many cases offset the slow speed of signal generation and transmission. When the signal from each channels arrives in the brain, it is transmitted to many other neural cells. As the signal moves from one set of neural cells to other set of neural cells, various acts of computations leading to processing signal are performed.

All the signals from the eye into the brain are in the form of digital pulses. Our flicker fusion rate is determined by the ability of the eye and brain to process the optical information. For us, this speed is about twenty images per second. It is a remarkable feat; our brain receives, processes, and understands millions of portions of an image in less than a few milliseconds including high-definition color interpretation with depth perception. From any modern communications standards, it is a truly remarkable act of information processing!

Computer science is the mathematical framework for processing binary information. Just like our rational mind, a computer also processes binary information. A computer takes certain binary information, in the form of an input, and by suitable manipulations converts it into an output. The manipulations are defined by mathematical algorithms and carried out by physical devices. Obviously the types of manipulations that can be carried out are limited by the types of physical devices available. Efficient computers are those that reliably carry out their tasks using the least amount of resources. Considering the living organisms to be specialized supercomputers, one can understand their abilities and efficiency in manipulating information.

Modern binary computers can trace their origin in humble beginnings as calculating machines. The earliest of calculating machines are the abacus.

The mechanical calculators came next. French mathematician Pascal in 1642 made a mechanical calculator that used the decimal system to perform addition and subtraction. George Boole in the early 1800s developed Boolean algebra. It led to the mathematical logic by which computer circuits are still designed. However, modern concept of the computer is based on a mathematical framework laid down by Alan Turing in 1936. Turing introduced us to a computing system that we know today as a programmable computer. Therefore, our modern computers are often called the "Turing Machine" in his honor. The mathematician Alonzo Church, Turing's doctoral thesis advisor, along with Turing generalized this concept into what is known as the Church-Turing hypothesis. This fundamental and pioneering work formed the foundation of entirely new field which is today known as computer science.

John Vincent Atanasoff, a theoretical physicist at Iowa State University in 1939, built the first electronic computer. Shortly then, Alan Turing and colleagues in Bletchley, England, designed and built a computer that could perform many mathematical calculations. Von Neumann and Turing hoped that computers could multiply our capacity to think, just as the capacity to perform work by our muscles had been multiplied by machines developed during the industrial revolution. Today, we are well on our way to realizing that dream.

Clearly, there are only a few architectural similarities between brains and computers, but there are a lot of similarities at informational processing level between the two. The brain comprises of billions of neurons with tens of thousands of connections per neuron all connected to facilitate parallel computation. The general architecture of parallel computational networks is similar to neurons within the brain. Using such architecture, one may be able to take advantage of simultaneous processing with cross-resolution of inconsistent concepts. Such connections have been modeled as "neural networks" by many researchers. These neural net models are based on rather simple assumptions on interneuronal synapses as switches between neurons. These models, as researchers have found, can simulate informational systems capable of learning, independent recognition, and even imagination and basic consciousness.

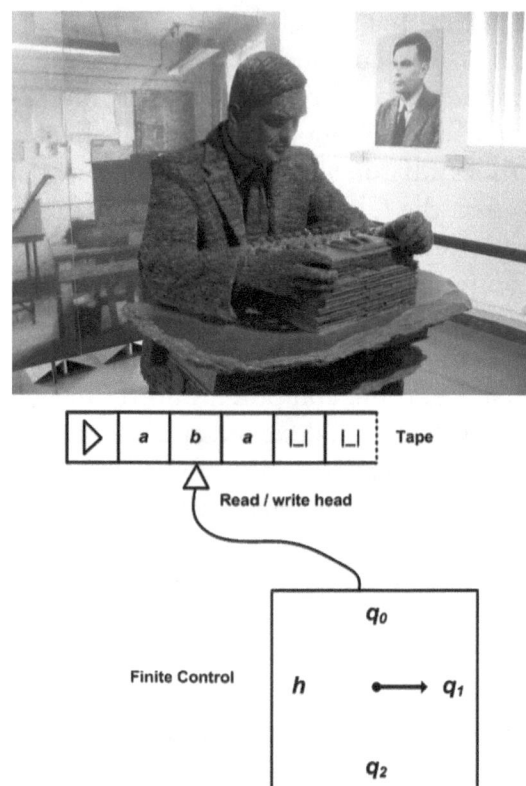

Modern concept of the computer is based on a mathematical framework laid down by Alan Turing in 1936. Turing introduced us to the notion of what we know as a programmable computer. Therefore, our modern computers are often called "The Turing Machine" in his honor.

Is our rational mind just a sophisticated, natural computer and a computer just an artificial rational mind from this computational point of view? The answer appears to be "yes" with one important difference. Our technology is evolving from binary—to quantum information processing; however, biological evolution originated from quantum information processing appears to head toward more evolved binary processors. As rational thinkers, humans have discovered the power of Internet by learning to connect binary computing systems in the form of a giant processor of binary information.

It is essential to recognize that the Universal generalization of Turing Machine is not simply an abstraction. It embodies a physical device subjected to the laws of physics. It is rooted in the classical laws of binary information. In 1985, the theoretical physicist David Deutsch devised a quantum analogue of the Universal Turing Machine. It is called the Universal Quantum Computer. It is again not an abstraction, but a physical device based on powerful laws of quantum information. He has argued that such universal quantum-computing device can simulate the behavior of any finite, physical system.

"With a head start of a billion years or so, life has devised human consciousness; human technology is catching up rapidly," remarked S. Hammeroff, who is an anesthesiologist and a renowned brain researcher. As a giant quantum-computing system, we are already networked as a part of a grand quantum-computing system, and now the local systems of binary computation are being evolved. In fact, all life is fundamentally rooted in quantum computation but evolving greater level of abilities to facilitate binary computations.

As far as we know, rational mind is the final developmental frontier. The neocortex is the crowning solution, the climax of informational system design the cosmic mind has produced. The neocortex continues to evolve rapidly. The neocortex uses the same basic structure and unit for computation. It has evolved so fast that the brain had to fold itself up to fit more of the computational matter into the skull. All human behaviors are intimately connected to it and may be the result of this evolving mind interfacing with emotional and cosmic mind. It is rooted in information. It has courted reason. It has pushed reason to its limits, the reasoned truth. But is it also the absolute truth?

## Reasoned Truth

Scientists use logic involving deduction and induction to reach conclusions which they believe as truth. A systems view leads us to believe that subsystems within our universe should be all connected, and, therefore, all the reasons as well as conclusion should be consistent with each other. Conclusions reached in this way are considered more certain than sensual perceptions as isolated or empirical experience.

At the heart of Buddhism is a similar concept that one thing or act or an event causes another. This idea of a causal connection or connecting through cause and effects is celebrated as dependent origination (or *paticcasamuppada*) in Buddhism. In other words, the cause or the origin of one thing or an act is dependent on what preceded it. And everything in the universe is connected to rest through a causal connection. The only thing that is unconditioned or "causeless" is the state of Nirvana. Hindus believe in the laws of Karma or action. It is also a reflection of belief in phenomenology of cause and effect. The Karmas connect all the present, past, and future actions or events through reason.

Is the truth resulting from the reasoning of cause and effect tenable and absolute?

The question concerning the role of reason in confirming absolute truth has been debatable. What types of first principles, or starting points of reasoning, can lead someone to come to true conclusions? If one believes in empiricism, only sensory impressions are available as starting points for reasoning and may be to finally reach truth. If one follows idealism, there is a "higher" reality, from which truth can be extracted without needing to rely only on the senses. This higher reality is, therefore, the primary source of truth.

Since the seventeenth century, rationalists have considered reason to be a subjective faculty. For Descartes, Leibniz, and others, this was associated with mathematics. Hegel believed that all reality is reason. According to him, whatever is real is rational, and whatever is rational is real.

The reality based on reason seems to have one universal need. It needs an entity with consciousness and ability to reason. Our rational mind provides that ability to reason. It, however, needs housing or a living body for its survival and fuctionality. The latest brain science is leading us to believe that living body is really an emotional mind or shelter of rational mind. Before we proceed further on that road to digital divine, let us explore the nature of this mind and how is it different than the rational mind.

———ᨆᨆ◦◦ᨆ◦◦———

# Chapter 7

# Emotional Mind

Expression of emotions on the face

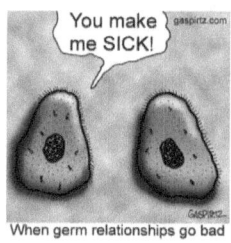

When germ relationships go bad

Emotional mind, common foundation in all life. The essence of life is not only matter, but it is information that is also at the root of life itself, a mind that creates and assembles information by supplying it a reason or utility.

## When Emotions Rule

Ruby was blessed. A teacher by profession, she derived pleasure in teaching young minds. With her small family, she lived in the suburb of Los Angeles, a quite unassuming life. Ruby was highly social and took pleasure in the company of friends and family. Her passion was to meet new people and bring people together. She is a good cook, and she used her cooking talents to feed her passion to make friends. She was particularly proud of her two young boys Atul and Aseem. She was close to Atul, her youngest. As a child Atul used to be quite sick, which was reason enough for him to be very close to her mother's heart. She spoiled him. Both parents did not expect much out of Atul, but he surprised every one with excellent grades. He was able to skip the sixth grade due to his excellent grasp of advanced subject matter. He scored the highest SAT score in his high school's history. He was named valedictorian. He studied math, physics, and focused on premed in college. In his medical school entry exams, Atul was ranked among the top 1 percentile.

Atul died in a train crash that occurred Friday, Sept. 12, 2008, in Chatsworth, California. He was heading toward Simi Valley, on his way home to see his parents, for the weekend when the Metrolink passenger train that he was traveling in collided with a freight train. Atul, only twenty years old, a brilliant physics and mathematics student planning to attend medical school, left this world with all his dreams unfulfilled. The family was devastated, especially Ruby; she stopped eating. Powerful emotions took hold of her personality. Her behavior changed. She stopped meeting people. Alone, she thought of attempting suicide several times. Finally, after many months of such incredible change in her personality, she is returning to some normal functions.

Why do emotions have such strong hold on our behavior? In his book *Emotional Intelligence*, David Goleman discusses this behavior in terms of emotional intelligence, as oppose to rational intelligence. Most of us are familiar with our rational mind. Is there a separate mind that is responsible for such intelligence? What are these emotions, and why do they affect us so powerfully? Is there a fundamental chemical, informational, or computational basis for our emotional mind? The answer comes from the new research in brain science, and it is yes. It turns out that our emotional intelligence comes from our emotional mind which processes information in a very different manner than our rational mind. This mind spans throughout our body and is impossible to separate as a separate physical entity. Therefore, our new understanding is leading us to a belief that our body is truly our emotional mind. Our rational mind lives inside our emotional mind whereas our emotional mind lives in the grand cosmic mind. Our emotional mind turns out to be an incredibly sophisticated information processor that mediates all communications between our rational mind and the grand cosmic mind.

## Conscious Matter, Life

Among all the works of the Cosmic Mind, the most fascinating is perhaps the "life." Even more so are us humans, who are pondering this fascination. So is our ability to ponder, to be fascinated, and to live this life. Although our universe is full of fascinating and complex systems and structures, life may be one of the most sophisticated structures that we know of.

Most matter can be easily classified as living or nonliving. Fish and birds are living matter while crystals and clouds are not. For some matters, such classification is not so easy. Viruses, for example, are one such borderline case. Biochemical soups of changing RNA strings in molecular genetics laboratories may be another. This indicates that life may not be a black-or-white property of matter, and some of gray do exist.

Living organisms share certain properties that are recognized almost universally with the trait of being alive. These are motion, organization, growth, reproduction, intelligence, and consciousness. Life is all around us. Incredible diversity of forms, ranging from microscopic bacteria to old giant trees, from simple algae to extraordinarily wide assortment of birds, from an incredible variety of fishes to astonishing thriving colonies of tube worms at deep-sea hydrothermal vents. No one knows how the first form of life emerged. However, it is often assumed that the first life-forms may have spontaneously arisen out of a prebiotic chemical soup. With time, self-replicating groups of molecules evolved and became more and more complicated. Their abilities to self-organize caused "cells" or cellular structures, the building block of life, to emerge. Cells changed further into roles that led them to harness energy from Sun, other sources, and reproduce and move about in their environment. This ability allowed them more control over their energy supply and physical conditions.

From the simple origins, life has evolved into a wide-ranging hierarchy of forms. Individual living entities live on as they exchange materials, energy, and information with their local environment. Different species of life exploit different niches in the environment. Initially, most energy may have come directly from the Sun through a chemical change. However, with time, fundamental changes took place. Predators evolved. Instead of getting their energy directly from the Sun, these creatures engulfed and digested the smaller and less powerful life forms. This chain stopped with the life forms that got their energy directly from the Sun or some other source of energy. We are acutely familiar with this arrangement of food. It is called the food chain. Most of the life, starting from single cells to large and complex humans, get their energy through this food chain and ultimately from the photons of the Sun. Humans are one of the most successful forms that life has produced. We are, without a doubt, now at the top of this food chain. Our quest to know and understand other life forms and ourselves continues. It is clear that we have made incredible progress in understanding various aspects of life. However, do we understand what actually life is?

About seventy years ago, Erwin Schrödinger tried to answer this question based on known laws of physics. His insight has since inspired many researchers to explore the molecular basis of a living organism. Early biochemists were disappointed when they realized that compositionally, life was not anything special. In other words, all the atoms that made life were widely available on Earth perhaps throughout the universe. Later, with the arrival of quantum physics, this material distinction lessened even further as all atoms are composed of quantum entities that occupy no space and are imaginary mathematical points or computational waves of potentialities in the digital ocean of existence. Like all other matter, life turned out to be composed of various combinations of quantum particles entangled and engaged in quantum computation at the smallest level. So what makes life behave so different from, let us say, nonlife? The distinction comes in the nature and quality of information processing that arises because of incredible organization, which spans from quantum to atomic to molecular and all the way to macro forms. So how does this larger organization or extraordinary computational abilities come about?

It turns out the whole universe had to conspire to produce such computationally astute structures. At the atomic level, most of the elements that engage in living process were made in the big bang. The hydrogen was the first element produced. The rest of the elements that make life were produced at the center of massive stars. All of these heavier elements were spewed out into space in some massive explosions of stars that may be several times the mass of our Sun, generally at the end of their life cycle. Repeated cycles of intense heating and cooling combined matter in myriad of chemical forms.

Today, it is possible in the laboratory conditions to take the common elements H, C, N, and O; stir them together with some heat, pressure, and electric sparks; and obtain molecules of life such as water, methane, ammonia, sugars, amino acids, nucleotide bases, and so on. In fact, today we have evidence that these molecules exist even in the interstellar clouds. In fact, it is not difficult to arrange these molecules in an orderly manner just like in a crystal or jumble them up in a random ensemble as in a gas. However, living organisms are neither ordered crystals nor random mixtures of their building blocks. The building blocks of a living organism are linked in a precise fashion to make functional parts with different levels of complexity. So far, we have not been able to produce a living cell in a laboratory. It does not, however, mean that we will never be able to do so. As we learn more about manipulating nanostructures and molecular structures, which are interacting at the quantum level, we will understand how to form and manipulate nonlocal links between building blocks, which are often indirect and not physical. A mind or information network is essential to manage this incredible informational order among the building blocks. This was the notion that was highlighted by Schrödinger in his decisive work.

The essence of life is not only matter, but it is information that is also at the root of life itself, a mind that creates and assembles information by supplying it a reason or utility. All life has some mental awareness, coordination, and connectivity. This relationship between life and mind raises another level of scrutiny. If we considered plants, bacteria, insects, and mammals, for example, each may disclose diverse mental activities, resulting in varied levels of behaviors including what we call as intelligence. Is there a basic mind that forms the foundation of mental connectivity among all forms of life? The answer is yes. It is the emotional mind. This emotional mind seems to have a common ancestry of informational molecules present in the simplest of single-celled living entities to complex human beings. New research in brain science suggests that it pervades throughout our living body. In fact, it is so intimately connected to all structures and functions of life itself that one may even say that our body is our emotional mind.

## Living Body or An Emotional Mind

Since breaking the genetic code in the early 1950s, cell biologists have favored genetic determinism, the notion that genes "control" biology. Virtually, all the cell's genes are contained in the nucleus. It is, therefore, was the first organelle to be considered as the "command center" of the cell, the cellular equivalent of the "brain."

Today we know this notion is false. We know that if brain or the command center were removed from a large animal, the interruption of physiological functions would lead to its death almost immediately. However, biologists found that in some experiments, enucleated cells survived for several months. These cells had no nucleus and yet continued to provide complex responses to environmental and cytoplasmic stimuli.

So where does the brain live for single-celled organisms? The new research is leading us to a novel class of mind, which is essentially similar in all life forms, whether single cellular or multicellular, even as highly advanced multicellular life such as human beings. It is leading us to the presence of highly sophisticated information network involving an incredible array of information molecules that forms the basis of this mind, which we will call as emotional mind. It is the brain of all life or at least one of them.

### Human Genome, DNA, Informational Blue Print of Life

The Human Genome Project began studies with a miniature worm. It was a primitive organism, barely visible with human eyes. It had a genome of about 24,000 genes. Then the researchers decided to complete one more genetic model before getting to the human. This was the fruit fly. Large amount of information was already available on the genetics and behavior of fruit flies. It would be a great learning experience before getting to the complexities of most advanced living entity, the humans. Of course, most

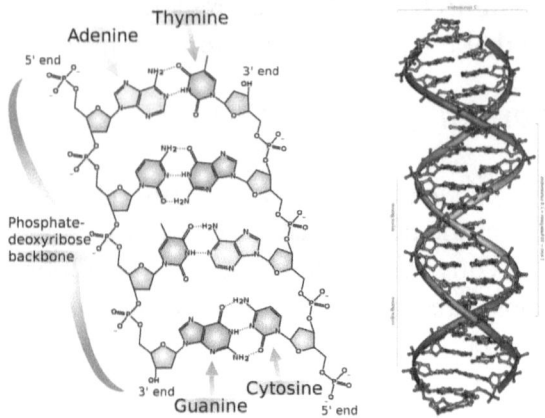

DNA Structure

*The human Genome is a digital program, when carried out; it produced an informational blueprint to make the human being. The incredible computationally astute structure of DNA molecule describes each individual is in a digital code. The backbone of DNA is based on a repeated pattern of two structures, first a sugar group and second, a phosphate group. The full name of DNA is deoxyribonucleic acid. The letter D stands for deoxyribose the sugar group. It is a modified form of ribose, another similar sugar group. The code is made up of four possible values. These are represented with four letter values: A, C, G, and T. These letters are named after the four complicated organic bases, namely cytosine (C), thymine (T), adenine (A) and guanine (G) that each of the DNA strand attaches itself to. These bases are always paired with their reciprocal base.*

researchers expected human genome to be large and complex. The fruit fly turned out to have about 18,000 genes in its genome. How could the primitive worm, with apparent simplicity, had 24,000 genes and this incredible, complex flying machine had only 18,000 genes? The red flag was already there, but researchers continued to characterize the human genome. The major shock, however, came in 2001 when scientists found that in the human genome there are only about 25,000 genes far below expectations of nearly 150,000 genes. Our human genome did not have enough genes to account for our own perceived complexities. It shook the foundational beliefs held dearly in classical biology as biologists struggled to get some answers.

The answers came from epigenetic. It was proteins, not the genes, which turned out to be at the core of human complexities. The human genome was not the command center; it was simply a digital program when carried out; it produced an informational blueprint to make the human being. In other words, the incredible computationally astute structure of DNA molecule describes each individual in a digital code. The backbone of DNA is based on a repeated pattern of two structures, first a sugar group and second, a phosphate group. The full name of DNA is deoxyribonucleic acid. The letter *D* stands for deoxyribose, the sugar group. It is a modified form of ribose, another similar sugar group. The letters N and A stand for nucleic acid that represent a set of information codes. These information codes are made up of four possible values. These are represented with four letter values: A, C, G, and T. These letters are named after the four complicated organic bases, namely cytosine (C), thymine (T), adenine (A), and guanine (G) that each of the DNA strand attaches itself to. These bases are always paired with their reciprocal base.

The human genome comprises of about 3 billion of such base pairs. The arrangement truly represents an information code or program. One can understand this complex genetic code through basic arrangements called codons consisting of three base pairs. Each of the base pairs may represent one of four possible values as described above. Together, these represent precisely sixty-four arrangements of various base pairs. These represent informational fingerprint of all life as we know it. Some of these sixty-four combinations are used to produce proteins from which the human organism or all life for that matter is made.

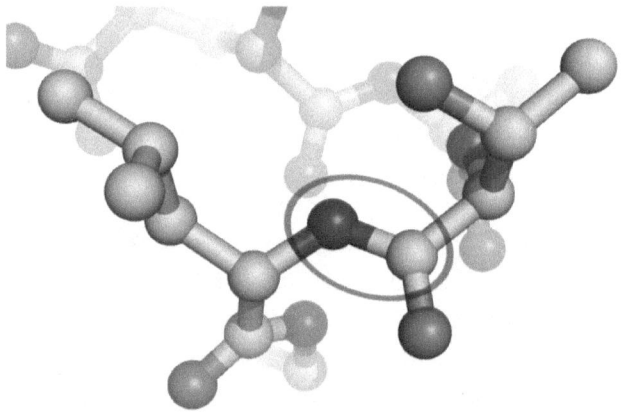

A peptide, molecule of information. Peptides have long been recognized as one of the first materials of life. Peptides are in reality proteins with one critical difference. The number of amino acids is usually less than two hundred.

Each protein can be described by a particular informational sequence of these codons found on chromosomes. As the program, in the genome, is read, the information dictating a particular construction of a protein is provided to the cell. In humans, more than one hundred thousand different proteins are made in cells from these informational sequences to form and maintain the overall human body. It is hard for most of us to imagine that such incredibly complex information processing produces all proteins ranging from simple to quite complex, some even totaling to thousands of precisely arranged groups of amino acids. If we could shrink ourselves to the size of one of these proteins, the cell would look like a highly sophisticated, modern factory. Precision, reproducibility, and lack of waste in each of the process carried out within the cell would outwit even the most sophisticated form of modern manufacturing scheme such as lean manufacturing or Toyota Production System.

**Proteins Or Information-Driven Architects of Living Structures**

All proteins are long chain polymer molecules with amino acids as repeat units or building blocks. There are twenty different types of amino acids. Each has a distinctive shape. The sequence of amino acids decides the shape and size of the final structure. Imagine each of the protein molecules as a self-knitting yarn. A yarn that has knitting instructions built inside its structure. Of course, we humans are far away from producing any such technology. The sequence of amino acid tells the protein string how to fold and form a specific structure. Changing the sequence of amino acids in the chain changes not only the final shape of the protein molecule but the way it knits into the whole structure. Amazing, isn't it?

What kind of structures do these protein molecules make? Many different kinds! The structure and composition of these macromolecules is such that it enables them to produce a large variety of biological structures. However, the most critical ones are the shapes almost like gears that engage with each other. James Berger, a biochemist and structural biologist who holds joint appointments with Berkeley Lab's Physical Biosciences Division and University of California Berkeley's Department of Molecular and Cell Biology, and Nathan Thomsen, a graduate student in his research group, have captured a critical action snapshot of an enzyme known as the Rho

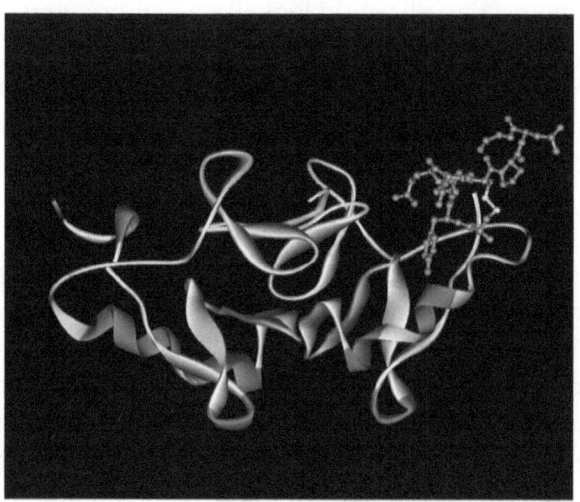

Proteins, the architect of informational structures in living beings. All proteins are long chain polymer molecules with amino acids as repeat units or building blocks.

transcription termination factor. In bacteria, the Rho motor protein binds to a specific region of messenger RNA and translocates along the chain to selectively terminate transcription at discrete points along the genome.

"We have shown that the Escherichia coli Rho transcription termination factor functions like a rotary engine, much like the motors found on certain classes of propeller airplanes," says Berger. "As the motor spins, fueled by the chemical energy in ATP nucleotides, it pulls RNA strands through its interior, an action that enables Rho to walk along RNA chains. Interestingly, the rotary firing order of the motor is biased so that the Rho protein can walk in only one direction along the RNA chain."

For most of us, it is hard to imagine that there is a mechanical gear system that is working at the molecular and nano level within our own bodies. In fact, all motions within a living being are a result of proteins gears being engaged by signals that turn on and off much like a lock-and-key mechanism.

Almost all activities that require muscle movements such as respiration and digestion are all results of these coupled-protein gears working in concert to create a specific function in the body. Almost all the functions of the cells owe their success to functionality of such protein gears. Making such gears is essential for maintenance and growth for cell and the whole organism. Once cell identifies a need to make gears for itself, it needs the blueprint from the DNA. A copy of the blueprint called RNA is then used to build the required proteins. Proteins not only provided mechanical structures and molecular gears that living structures could depend on, they turned out to be the critical molecular building block of the cellular command center and information network for all life.

Each of our one hundred thousand proteins has a specific length and sequence of amino acids that control how this protein knits into a shape. Along the length of the protein chain, there are binding sites. These are often due to amino acids that have either positive or negative charges. The charges provide sites where various kinds of chemical signals can bind. What is bound to these sites decides the final shape of the protein or the structure it is part of. Environmental signals such as hormones, drugs, and other biochemical factors can latch on to the binding sites and thus cause

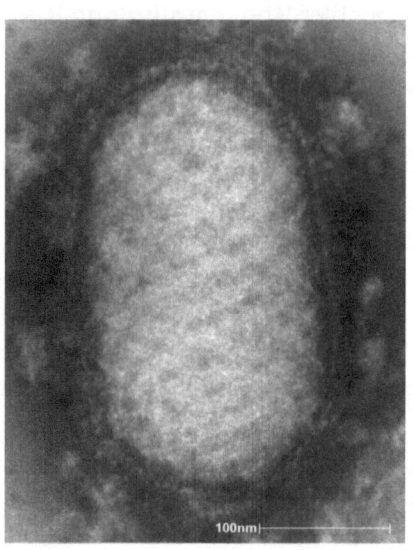

100nm|————————|

A single living cell, incredible computational, informational entity. Although our universe is full of fascinating and complex systems and structures, life may be one of the most sophisticated structures that we know of.

the protein molecule to "adjust" its shape. As the protein adjusts its shape, the structure it is a part of moves. Such cooperative movements lead to all motions or movement in all living entities from primitive cell to complex humans. How fascinating!

Of course, lock-and-key analogy is a gross simplification of what the real process may be. In reality, the protein gears are vibrating and trying to assume a preferred conformation. Since these dynamic states are rooted in quantum computations, the conformational states may be impacted by both local and nonlocal variables. Local variables such as chemical signals were among the first to be recognized by biologists. The approach has been so successful that a completely new pharmaceutical industry was founded to make these chemical signals called drugs. Most of us are far too familiar with these chemicals than frankly we want to! If you experience a disease, chances are your protein gears are not engaging correctly. A suitable drug signal may change the protein gears curing one from a diseased state into health.

Quantum mechanical "signals" are invisible and informational supplementing the local physical or chemical signals. In fact, the invisible quantum field signals may be order of magnitude more efficient than drugs or chemical signals. Many Asian medicine, healing practices, for example, acupuncture and meditation are based on consciousness of the human attention to these quantum fields.

On the surface of any living cell at any one time, there are over a hundred thousand different protein switches. These switches respond to a massive variety of environmental signals in a cooperative manner. If we want to know what the cell is doing, we cannot look at any one signal or switch; we have to identify what bunches of hundred thousand switches are doing collectively. They collectively control all functions of our lives through awareness and response to the environment. These membrane switches are fundamental units of our perception as they sense environmental inputs and allow us to respond to the environmental needs. These informational switches turned out to be peptides, the molecules of information, or you may call them "baby proteins."

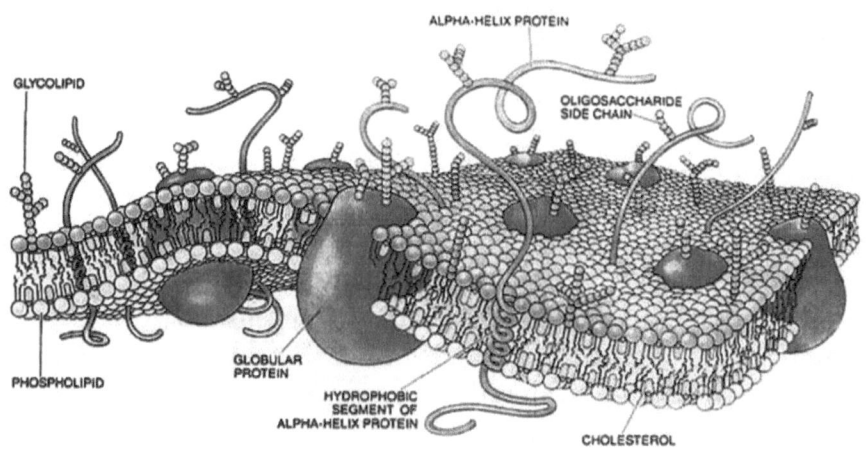

GLYCOLIPID

ALPHA-HELIX PROTEIN

OLIGOSACCHARIDE
SIDE CHAIN

GLOBULAR
PROTEIN

PHOSPHOLIPID

HYDROPHOBIC
SEGMENT OF
ALPHA-HELIX PROTEIN

CHOLESTEROL

On the surface of any living cell at any one time,
there are over a hundred thousand different protein
switches. These switches respond to a massive variety
of environmental signals in a cooperative manner.

## Baby Proteins or Peptides, Molecular Sensors of Information

Peptides have long been recognized as one of the first materials of life. What are peptides? Peptides are in reality proteins with one critical difference. The number of amino acids is usually less than two hundred. Now, if you recall, a protein could have thousands of amino acids in each molecule; there is really no limit.

The process of protein assembly can be compared to putting together information in a language. The amino acids are the letters of this language, and peptides along with proteins are the words of this molecular language or simply communication signals. The information network that these signals form directs every cell, organ, and system in your body!

Oxytocin was the first peptide ever to be synthesized outside the human body. Oxytocin plays a critical role in childbirth. For example, it produces the contractions of sexual orgasms in females, which eventually work to expel the baby. In addition, it acts to produce maternal instincts. This unifying role of peptides, coordinating physiology, behavior, and emotion seems to be common for humans and other animals.

Candice Pert, a neuroscientist and psycho pharmacologist, in her book *Molecules of Emotions*, describes certain peptides primarily used by neural network as the molecules, which affect the emotional state of the living being in a significant manner. She called these molecules as the molecules of emotions. Since these peptides are used by the neural cells, they are also called "neuropeptide." In the early 1970s, Candace Pert discovered the "neuropeptide" opiate receptor site on cells. This led to many vital advances in developing the scientific understanding of the mind-body connection. What she describes as molecules of emotion are critical parts of the information network that forms the basis of our emotional intelligence or mind.

Neuropeptides were first thought to be produced only in the brain. Now we know that almost every cell or tissue in the body produces neuropeptides. Over two hundred different neuropeptides have been discovered to date. The first of which were hormones. Neuropeptides are chemical messengers

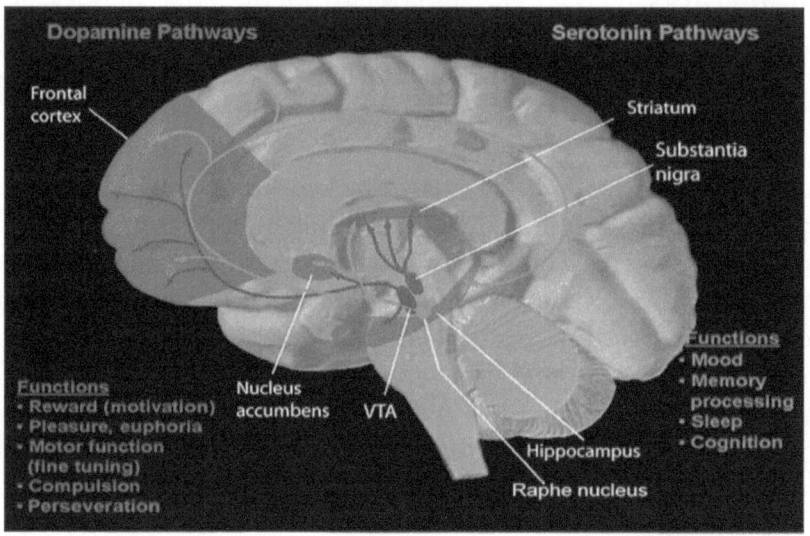

Emotional circuits or flow of molecules of emotions in human brain.

that carry messages to receptor sites on cell membranes throughout the body. How do they do that? Simple, neuropeptides circulate throughout the body by floating in body fluids such as the blood, extracellular fluid, and cerebrospinal fluid. The average cell may have literally thousands of receptor sites for neuropeptides, which are constantly opening and closing. But that is nothing; main action is in the brain cells which may literally have millions of these receptors!

Incredible amount of information processing goes on much like a lock-and-key mechanism or binary bit switch where the neuropeptide is the "key" that opens the "lock" on the cell membrane. The switch is turned on or off. When it is on, it causes physical and informational changes in the cells it lock onto. However, as described before, "key fitting into a lock" analogy is too static an image for this incredibly dynamic process. Pert uses the description of two voices or sounds. As peptides and receptor site produce the same note, the signal is turned on. According to her, the process is similar to peptide trying to ring the doorbell of the cell to open it. Clearly, the true mechanism is far more complex and has its roots in quantum information processing, an interface where binary computation synchronizes with the quantum computation systems of the cosmic mind. This is also an area of incredible current scientific scrutiny.

### Emotional Mind or Symphony of Floating Molecular Information Network

All systems, from simple cells to tissues to complex organs of the body, constantly exchange neuropeptide and thus exchange emotional information. It is a dynamic process, which involves the whole body as a complex information network. The internal feelings states such as wants, attention, intentions, feeling good, angry, or depressed all arise from such exchanges. In fact, emotional personalities arise from behaviors that induce such repeated responses involving the neuropeptide exchanges. Every change in mental-emotional state causes changes in the body physiology leading to corresponding changes in the mental-emotional state of beings. This is clear body-mind connection facilitated by the informational molecules.

Let us examine some of the scales to describe observation of the universe at the smallest level to the largest level. Our eyes can observe only so much. Smaller objects require microscopes (SEM, STM, etc.) whereas larger objects require telescopes. Forces can be very large (big bang) to very tiny operating at quantum level.

Human system has amazing capabilities to receive, analyze, and respond to data inputs from very tiny to large. The force called feeling measures a range from extreme empathy, softness, and love to overpowering others to inflict pain.

—Anonymous

According to Dr. Pert, the collective activity of these biochemical messengers or information molecules displays emotional intelligence of the organism. This intelligence is communicated in the form of information. This information transfer takes place over a mobile molecular network linking all of our systems and organs, engaging all of our molecules of emotion, as the means of communication. Through the action of such molecules, Dr. Pert concludes that our thought and awareness can actually transform into the matter and may transform our physical body.

Deepak Chopra, a physician and renowned author of several books on body mind connection, stated, "The discovery of neuropeptides was significant because it showed the body is fluid enough to match the mind. Thanks to messenger molecules such as neuropeptides, events that seem totally unconnected—such as a thought and a bodily reaction—are now seen to be consistent. The neuropeptide is not a thought but it moves with thought, serving as a point of transformation. A neuropeptide springs into existence at the touch of a thought, but where does it spring from? For example, a thought of fear and the neurochemical that is turns into are connected in a process that involves a transformation of nonmatter into matter."

Our mind, spirit, and emotions are unified within our physical body in one "conscious information system," where informational molecules such as the neuropeptides and their receptors form the foundation of our awareness, manifesting themselves as our emotions and beliefs. They deeply influence how we response to and experience of the world. This vital connection between mind, body, and emotions facilitates the informational molecular language that allows communication within all biological systems. What we see is an image of an emotional mind, one that is molecular, mobile, and moves throughout our body, located in all parts of our body and not just in the head. This bodywide information network is informationally ever-changing dynamic and amazingly flexible. It is an incredibly astute information network, directing and admitting information simultaneously, intelligently guiding what we call life, health, and wellness.

Erwin Schrödinger's assertion that life is an interplay of energy, entropy, and information is perhaps more true today than seventy years ago when he first proposed it. Systems for information processing have been evolving within biological life-forms ever since life emerged. The most evolved information-processing system currently appears to be the life itself. The molecules of emotions or information molecular network, which appear to be universal in all living beings, form the foundation of emotional mind. From single-cell organisms to most advanced living system such as humans, all perform simple to complex tasks involving rudimentary decision making, behavior, and organization using the same information network of these molecules. This provides us with a new view of emotional mind which involves computational as well as holistic nature of biological organisms. It is becoming clear that it is not possible to separate the body from emotional mind, in humans and in all living beings, as all aspects of cell and organisms functionalities are rooted in computation and information processing. The living body is, in fact, the emotional mind.

This incredible creation of the cosmic mind is the basis of all living entities, a powerful information processor, which controls all life. It is informational, computational and reward based. The feeling good is at the core of its control mechanism. Forming relationships, associations or connectivity with in as well as with others is at the core of its operations. It is computationally astute. It combines quantum information processing with binary processor to maximize computational efficiencies leading to the formation, survival and growth of the form. It provides housing and connectivity for rational intelligence. It provides an interface that computes, processes information, and is rooted in reason on one hand, just like the rational mind. However, on the other hand, it provides connectivity with its quantum roots, to the cosmic mind. Here, is intimately rooted in quantum state of superposition, which is *one* for all. In this mode, its reasons and intelligence become different. They are guided by the nature and reasoning of none other than the grand cosmic mind or digital divine. Its ultimate reason becomes love, compassion, and empathy.

Hurt and helpless, Ruby blamed first herself, then God, for the death of her son. No amount of rational reasoning could console her. It was as if a complete chapter from emotional book of her life was torned. She felt as if a piece of her living heart split from her being. She lost the connection to the cosmic mind, at least temporarily, and lost her will to live. Her old emotional circuitries challenged and perhaps rewired. Her search for new ones continues, the ones that can provide her respite and perhaps meaning of it all.

# Nature of
# Nonlocal Mind

# Chapter 8

# Computational Universe
## or
## a Cosmic Mind

Nature isn't classical, dammit, and if you want to make a simulation of Nature, you'd better make it quantum mechanical, and by golly it's a wonderful problem!

—Richard Feynman, 1981

# Road To Digital Divine

## Digital Nature of Our Universe

There is rapidly growing consensus among scientists about the digital nature of our universe. The scientist John Archibald Wheeler claimed that, fundamentally, atoms are made up of qubits of information. As he puts it, "Its are from bits." He elaborated, "Every it—every particle, every field of force, even the space-time itself—draws its function, its meaning, its very existence from binary choices, bits or more fittingly qubits." What we call reality arises from posing yes/no questions. In a world made up of bits and qubits, the universe itself is the supreme computer and calculations seem like a divine act. It takes the choice between yes and no, the state of 1 or 0 or the most fundamental state of existence: I am/or not.

"All creation evolves from this irreducible or wholesome foundation. Every galaxy, every star, the smallest ant, each thought in our mind, each flight of a ball are but a web of elemental yes/no woven together," writes Ross Rhodes, a quantum physicist.

Qubit is widely seen as the tiniest constituent of existence. This new quantum informational element is seen to be the basis not only of matter but of energy, motion, mind, and life. If information theory holds, all movements, energy, gravity, dark matter, and antimatter can all be explained by elaborate programs of binary and quantum bits.

"But I do not make yes/no choices all the time," you may ask. All the actions that we do, as a human entity, are not made in the language of 0, 1, or yes/no. It is true that human actions rarely take place in the language of yes or no; however, each of the action are similar to programming in the higher level languages like Fortran, Pascal, or Basic. Just as a programmer, using these languages, is able to do tasks such as programming faster than if it were to be performed in binary language, we are also able to program our and other's action far more efficiently than if we started to program these actions at atomic or quantum level in the language of binary or quantum bits. However, whatever language we use, whatever action we perform is all reducible to the series of 0, 1, or more precisely qubit language. In fact, all the higher-level organization such as molecules, cell, tissues, organs

Quantum computers can evolve a
superposition of quantum states—each
could follow coherently distinct computational
paths till measure final output. Such "quantum
parallelism" could potentially outstrip power of
classical computers.

—David Deutsch, 1985

enable the shuffling of bits or qubits far more rapidly than programming at the quantum level, using individual yes/no programming. From this perspective, all actions and events in this universe are computational and informational, involving manipulation of these qubits; the input being the information, whereas, the output is the result of computation displayed in form of information order, structure, and entropy.

If the universe is a computer, where is it running? David Deutsch explains, "The universality of computation is the most profound thing in the universe." Since computation is absolutely independent of the "hardware" it runs on, studying it can tell us nothing about the nature or existence of that platform. Deutsch concludes it does not exist: "The universe is not a program running somewhere else. It is a universal computer, and there is nothing outside of it."

We are linked to one another, all beings alive and inert, because we share, as John Wheeler said, "at the bottom—at a very deep bottom, mostly—an immaterial source." This quantum computation of these qubits is a precise, definable, yet invisible process that is immaterial yet produces matter.

## The Whole or Cosmic Mind or Digital Divine

### Many Ways To Describe The Same

Most scientists may not believe in an old man with beard with a cosmic mind called God hidden in the inner working of the universe and controlling all there is, but most of them do believe that nature as a whole is a computational platform or an informational entity that is constantly taking measurements of, and through all the entities, that exist within its being. The rules of quantum information processing are leading scientists closer to the laws that govern this universal entity. The old laws of physics focusing on the matter and energy are being replaced by newer laws of information with much broader reach. Just like the uncertainty principle, the laws related to conservation of mass and energy, or quantum gravity, may all have roots in information. This almost-hidden scientific revolution is promising to take humanity from the "binary information age" to the "quantum information age." The underline core belief is that our universe is an informational entity or informational "whole" based on quantum information. For the sake of simplicity, I would call this informational entity as *whole* signifying that it contains all quantum information that makes our universe and there exists nothing beyond it. Now you may ask an age-old question. Is there a grand cosmic mind hidden in *whole* carrying out grand invisible quantum computations? Or you may ask, is cosmic mind a better way of describing this grand computational entity I have called *whole*? Or one may ask does *whole* or cosmic mind simply represent two different perspectives of describing the same entity? Along the same line, one may also suggest that the nature as a *whole* is a mind, and we along with our visible universe live in it. This is a familiar description, preferred by many religious doctrines and mystic beliefs about our universe. It may also lead to the suggestion that the *whole* is a digital version of divine or simply God. Are these simply many ways to describe the same?

From a scientific perspective, how close this informational, computational description of our universe comes to implying that a cosmic mind or God exists? Clearly, a description involving mind implies a lot more than science can infer about the nature of the informational, computational entity called *whole*, especially at this stage of development. Just the use of the word *mind* can trigger a lively philosophical and scientific debate. A mind is widely recognized as the seat of perception, self-consciousness, and thinking process; root of belief, memorizing, hope, desire, free will, judgment, reasoning, etc. These characteristics are normally connected to a human mind. Does cosmic mind share some or all of these characteristics? Or are there other characteristics that cosmic mind takes, which are unique to the grand cosmic mind and not present in a human mind? Alternatively, are there characteristics in the human mind that are not present in cosmic mind?

This is where a divide exists between science and religions or mystic, an objective versus subjective mode of inquiry. The science of quantum computation is in its infancy, and a comprehensive theory of quantum information processing is still in works. However, there is compelling evidence that the devices using the laws of quantum information, we will have Godlike powers and capability unimaginable to us with today's technology. Another important development that is crucial in bridging the gap between science and spiritual know-how is the recognition of the observer. The observer has become central to describing reality in framework of quantum mechanics. Spiritual virtuosos have always given prime significance to observer or subjective experience of reality. This is enabling the latest advances in quantum computation to further bridge the gap between subjective and objective mode of inquiry. Without a doubt, science is getting closer to providing foundational support to incredible body of subjective data about the cosmic mind; however, today it is not advanced enough to explain it all. I will take on this in details in the next chapter where we will talk about immortal and how subjective modes are able to access this layer of universal code.

Are we at this stage to be able to combine the two mode of inquiry and form one integrated picture? The answer, in my view, is a resounding yes. The progress in each mode of inquiry has been incredible, and we should at least begin to make efforts to bridge the gaps between the two. Cosmic mind, God or Digital Divine in my view, still may be appropriate descriptions of the informational, computational entity *whole* as these allow us to take steps to integrate the foundations implied from advancements in quantum computations to the incredibly huge body of religious and mystic data obtained through subjective mode of inquiry. From this point of view, the use of the terms, *cosmic mind, digital divine, God,* and *whole* especially in this book series are simply many ways to describe the same representing appropriate context and perspective. Human mind, in my view, is a great source of learning about the cosmic mind as it is made in the image of the cosmic mind. It, in fact, forms a bridge to subjective and objective mode of inquiry. When one embarks on a journey that combines these modes of inquiry, one arrives at a new model or perspective to understand basic nature of our universe or cosmic mind, a road that leads to digital, informational and computational divine. Accordingly, the cosmic mind or digital divine has the following two aspects of being:

1. Nonphysical Being or Abstract Reality
2. Physical Being or Visible Universal Reality

Both these aspects are present in the human mind. Our intelligence in the form of thoughts, perception, memory, emotions, equations, and imaginations has abstract reality associated with it. Physical reality of our mind is equally significant referring to neural network cells organization, coordinated motion of information molecules, and physical organization within cells, tissues, organs, and organism. Understanding the universe as a system or *whole* involves understanding both these aspects of its being and their emergent properties.

## 1. Nonphysical or Abstract Reality of the Cosmic Mind

Scientists know that quantum entities exist, and all of them follow one basic structure of quantum information. It is one of the most incredible simplicity that unites our existence. However, they do not know what leads to such grand simplicity. As I was researching this area, I came across this quote from Hindu holy book, Bhagavad Gita.

"And undivided, yet He exists as if divided in beings"
(Bhagavad Gita, ch. 13, verse 17):

Here, Lord Krishna describes the nature of *divinity* itself, in terms of quantum information. It gives one an indication that even though objective and subjective mode of searching for the fundamental nature of beings have been extremely different, but they tend to arrive at the same fundamental reality of existence. If there is *one* simplicity the cosmic mind can be reduced to, it is not a device or smallest particle of matter or a superstring or membrane of energy, as many of us may have suspected. The greatest simplicity of this mind exists at the abstract operational level. It is the manner in which all that exists in this mind computes, functions, or operates. According to Gita, it is universal and one may call it *divine* mode or true nature of being. It arises from two basic tendencies that fundamentally make the universal, cosmic mind itself. These are the following:

1. Tendency for One to Be Many
2. Tendency for Many to Be One

The simplest tendency of being many is a tendency to be two. In other words, one whole could be in one of two possible states, for example, a set of yes or no. Either exists or does not exist. Either yes or no, zero or one—a binary choice. The choice is always one or the other, never both.

The simplest tendency of *many* being *one* is the state of existence where both of the above binary choices yes and no exist together at the same time. One may interpret this as a choice that involves emptiness or zero influence, or one may say a state of inaction or utter and complete contentment or acceptance of many into one. So one may say that this algorithm involves the following states of computation:

1. Computational element or a system of computational elements leading to occupying one of the two positions on/off
2. Each computational element capable of acting as separate binary unit, when observed.
3. Each computational element capable of acting as part of a *whole* operating as *one*, when not observed.

This is also the basic structure of quantum information, and amazingly one may arrive at that from the thousands-year-old quote found in form of holy text. Smallest unit of quantum information, a quantum switch or quantum bit or qubit, represents such an algorithm. The switch takes one or the other position when interrogated about its state. Else, it stays in a surperpositional, wholesome, or state of oneness. This state of oneness is a state where parallel computations are taking place. All qubits in this state act as one computational entity or one mind. This is the strangest, most amazing state of existence found in nature as unmanifest or silent or empty state of being. It is like a common community mind, common to all qubits. Each qubit derives its direction to turn on or off from this common mental resource. Since all informational entities are ultimately based on qubit, they all derive their direction from cosmic mind. It is the essential nature of the cosmic mind.

Binary switch or binary bit is a special case of qubit. It derives its directions from association with other binary or quantum bits. An informational entity that comprises of all binary bits or bits entangled with each other gives rise to local minds such as emotional and rational minds. As emotional mind provides an informational interface that separates an informational entity from cosmic mind, the rational mind provides added computational capabilities to address local variables. This rational mind has similarities to binary computers as we saw earlier. The behaviors of an informational entity dominated by a rational mind will be remarkably different from the behaviors of an entity dominated by cues from cosmic mind. A combination of these two modes of computation leads to behaviors of all informational entities that occupy our universe.

From a reductionist point of view, the *whole* or cosmic mind, at the root, is a collection of quantum particles which exist in a computational state, or state of entanglement, or superposition state. Another way to describe it is as an ocean of digital waves of possibilities with no medium and no energy. These digital waves are the information waves that arise from mathematical formula only. The formula is like the software that operates the cosmic mind. You may also describe the *whole* as an ocean of silence filled with these digital waves of potential action or dissatisfaction. Or you may say, this ocean represents grand silence, which is characterized by "energylessness" or utter self-satisfaction or contentment. Here all dualities coexist in superpositional state as one. The infinity and emptiness or zero merge here. The state of death and alive merge here. The state of being and nonbeing coexist or merge in oneness.

The separateness coexists with oneness. Randomness coexists with utter and supreme reasoning. Wave and particles states coexist. The finite and infinite coexist. This is the state that all meditator aim to achieve. It is also the state of enlightenment, portal to enter the Grand Cosmic Information Processor. In this state of existence, the *whole* computes using mathematics as universal software, producing all physical exhibitions as physical expressions of mathematical formula. An amazing mathematical formulation that describes the universe beyond space, time, and with perfection; from smallest to largest, fastest to slowest, within and beyond physical boundaries, just absolute *truth* is contained in this realm of existence.

All the digital waves represent the superposition states of individual qubits or informational entities that exist in cosmic mind. In this state of superposition, the ocean of trillions and trillions of qubits act as one *grand* mind or *mega* quantum information processor. An unimaginably powerful quantum computational entity that produces and manipulates all there is in this digital ocean of existence. It is the omnipotent, omniscient and omnipresent universal, cosmic mind or digital divine. It is the creator and solution provider to all its problems, structures, and inhabitants. It is the governing intelligence, creator of all that is visible and invisible including us (*self*) as observers.

All calculations within *whole* are performed instantaneously in the nonlocal regime almost like a central processing unit or CPU, regardless of the location. The result of such calculations is the wave of guidance that pervades the whole universe. All results are communicated to all in an instantaneous manner. The essence of nonlocality is unmediated action at a distance. This nonlocal form of interaction jumps from one entity to another entity without interacting with anything in between. It is like as if the cosmic mind is the conductor of a grand orchestra, a dancer or informational Natraj. By sending the digital waves of guidance to each inhabitant in the digital ocean of potentiality, the cosmic mind directs all acts inside the grand mind.

Eventually the *whole* is one entity incorporating the smallest of the very small and the largest of the very large. Zero and infinity, quantum or binary computation systems, physical or abstract dimensions all merge in *one*. At the tiniest level, the physical devices that obey quantum-computing algorithm are particles such as electrons, photons, protons, and quarks. Even atoms, molecules and all other larger molecular devices follow quantum computing and qubit structure of information as long as they stay unobserved. What happens when an observer arrives at the scene?

## 2. Visible Universe: Observer Based Physical Reality of the Cosmic Mind

Our visible reality emerges as a result of our observation of the abstract reality residing in the cosmic mind. With a local mind, we observe and model the physical reality. Without a local mind, all there will be random collections of bits, bytes, and qubits. In fact, we coconspire to create a physical reality with cosmic mind. The quantum switches and binary switches interact with each other to form more complex network of information processing entities giving rise to the computational, digital nature of our visible universe. These entities stay in an abstract or unmanifest state participating in incessant parallel quantum computation, in the state of superposition, until observed.

The most prevalent speed in the universe is infinity. Most prevalent state is the massless, energyless, and timeless state of being. Speed of light is the dividing line that separates eternity from our entropic mortal reality. This is, in fact, the last stop for eternity, perfection, or timelessness. In this realm of our physical universe where mass and energy are finite, the time comes into being. Devices at tiny quantum level spend most of the time in

Our visible universe is a screen where all the energetic entities flicker in and out; these flickers together form shapes motions that are only observed by authorized viewer. Even though this movie is viewed or edited by the viewer, the primary creator, editor or director is always a part of the unmanifest. Because our viewing speed is slow, we miss most of the action. Because our comprehension is poor, we miss most of the understanding. In unobserved state, universe acts as one grand informational entity, within which all are interrelated, and all computations go on simultaneously.

—Anonymous

surperpositional state computing as interconnected computing elements of this grand quantum computer. In terms of information, small and simple informational entities combine to form larger, more complex informational entities or computational elements. As the molecules or even larger entities form, their quantum nature is preserved but due to larger size and greater visibility, these entities spend more time doing binary computation than quantum computation. Nanoscale devices truly struggle between classical binary state and quantum state. As the size goes up and particle remains single, it will spend more time in binary states.

Complex particles, such as life, involve quantum foundations in a larger classical body. These particles are complex computers carrying out computations that switch from binary to quantum modes. The living organism, in other words, is computationally astute and exploits both form of computation to aid its survival and living state. Humans are large, complex systems of such particles. We are made up of trillions and trillions of cells organized in a sac that is filled with informational molecules floating in our body fluid, all acting in concert as an incredibly complex computational system leading to formation of emotional personalities. We also compute from binary to the quantum level using our emotional and rational mind. As described in earlier chapters, a simple picture of life emerges. Our bodies are our, in fact, emotional mind. The universe is a cosmic mind; we live in this mind. Therefore, our emotional mind lives in cosmic mind. The two minds are informationally and computationally synchronized. This is true of all life, simple single-celled or advanced systems such as us humans. Since our rational mind lives in our emotional mind, one may say that our existence is truly a tale of three minds, cosmic mind, emotional mind, and rational mind. All three connected at local and nonlocal level and working in concert at all instants, weaved in tangled hierarchy to produce the consciousness, experience, and behaviors we have in our daily lives. An observation by such a conscious entity, directly indirectly, or through its intentionality, produces all physical and visible reality. What is the form of this reality?

The form in which such nature manifests itself is very familiar to us. It is our perceptual reality. It has an appearance of a mega distribution system. What does it distribute? At the most basic level, it distributes quantum information. All the natural systems visible to us, such as water cycle on earth, forest, food chain and all ecosystems such as pond can be understood as distribution systems, operating under certain rules and providing needed resources to their members. All these systems ultimately arise from this grand quantum information distribution system, distributing information, potentialities, and energy at the core level. Once these potentialities are delivered to a user, they are actualized to fulfill the needs and desire of the members of this distribution system. So one, at even more fundamental level can say that this distribution system delivers fulfillment of needs, desires, or intentionality of its members. It provides free will to some, and others exist in a state of complete surrender. Some others such as humans exist in a state somewhere in the middle of the two.

Internet as we have it today provides a relevant analogy. At the most basic level, Internet distributes information coded as zeros (0) and ones (1). Today we process incredible size of data on the Internet, and the Internet is an essential element of our lives. Internet is available on our desktop computers as well as on our mobile devices. Internet has come a long way from what it used to be in the past decade or so. But from its future potential, it is young and in infancy stages. It would undergo a series of evolution as technology advances. Recently in a keynote speech, Cisco System's chief technology officer (CTO) Padmasree Warrior described the future of the Internet and mobility to be synonymous, which means that we would have Internet wherever we need it. We are already quite close to that goal.

At the quantum level, our universe represents an incredible wireless mega-Internet working on quantum principles. This means that this mega computing system is like the Internet, but that is based on quantum bits rather than binary bits. The quantum Internet connects all the informational entities in this grand distribution system all the time irrespective of their location or mobility. We are always connected to it. In fact there is no way to bypass this connectivity.

This distribution system is powered by incredibly powerful quantum-computing core that arises because of various entangled qubits. Humans are still quite far from developing any quantum computer that is of any significant computing capacity. There have been some successes in developing working quantum computers with a small number of qubits. There are a number of significant problems that arise from influence of external systems or decoherance of qubits as the information content of such systems evolves.

The big bang provides ideal conditions to make quantum computers as there is no external system influence that could decohere the evolving quantum-computing system. As the quantum particles evolve and entangle with each other to form entangled superpositional state of a perfect quantum computer enabled by energyless, nonlocal communication at the infinite speed. The quantum computer is like wireless network, and every physical particle is a receiver of the nonlocal information waves that travel at the speed close to infinity. As the signals are received just like today's wireless devices as cellular phone or PDA, they get processed by our bodies. There are equivalents to much familiar Internet hardware and software. For example, to view the content of Internet, a browser program is needed. The browser to view universal quantum Internet exists within us. We can receive and transmit information to this quantum information network. There are security equivalents as well. Most of the humans are not allowed to intrude freely into this network. However, we will talk about some enlightened individuals who have been able to gain entry into this network later in this book series.

At the macro physical level, the ocean provides another relevant analogy of a distribution system where food and chemical are transported to various inhabitants. Ocean current carry nutrients, food, and other needed ingredients from one area of the need to the other. The digital ocean of *whole* is the ocean of potentialities. The waves are the unobserved mathematical, informational waves that travel at speed faster than the speed of light, transmitting nonlocal information, almost instantaneously and without involvement of any energy. The receivers are the islands of the observed structure or visible universe. The quantum distribution system is highly detailed and distributes information or potential information from the tiniest to the largest. It is the ultimate *truth* at all levels.

This essentially means that all the needs and desires live in the superposition state until a decision is made to accept and actualize by its members. Since this highly interactive distribution system operates at infinite speed, or instantaneously due to nonlocal transfer of information, the distribution system is always in present, irrespective of the speed of observation. The infinite speed of communication applies only beyond space-time. In this realm, eternity rules; fullness, completeness, and perfection are the rules.

Inside of me there exists a sea of potentialities, formless, timeless. It constantly provides me answers for my mind to accept. The sea of potentiality is beyond space-time and hence exists everywhere all the time.

—Anonymous

There is nothing that is somewhere but not everywhere. In other words, it is everywhere. This is terribly hard to understand by most of us.

However, this nonlocal reality exists as information waves, and we are entangled together with the system of such waves. Our body or emotional mind is able to feel these waves of guidance. Trillions of informational molecules and nano-size informational entities to macro level informational entities within our bodies depend on this communication for performance. This is the true entry point or mystics. Subjectivity is the only way to access this reality. In fact, objectivity is denied entrance to this portal. So if you are waiting for science to penetrate this barrier, you will have a very long wait, and the wait may never end. There is no other way for humans to access this reality other than going within and feeling it. I would highly recommend you not to wait. Why? It is because this reality is intensely sweet. There is no experience like this. Reaching here is the goal of this book series, and it has been the goal of humanity all throughout its existence, in my humble opinion.

# Chapter 9

# Seeking
# Divine

# Mortality, Immortality and Immortal

As pleasant, uplifting, and freeing immortality appears, our reality is that we eventually die. Death is one of the few certainties in human lives. All living beings die. Plants, animals and most others die, but they may not be aware of it. Humans may be the only ones that are aware of it. Most of us believe that this awareness is at the root of all spiritual inquiry. As we get closer to death, our search for what is beyond death intensifies. Why do we die? How can we live longer? What is beyond death? What is it like to be immortal?

Because of advances in medicine and better standards of living, humans are living longer. The average life expectancy for humans has significantly improved over the last few hundred years. Living longer is the obsession of current modern science. Death, however, continues to be a mystery. Most of us are confused on the issue of what happens to us after we die. Since no one has ever come back from dead, much speculation exists on this after-death state of our being, if at all there is anything like that. In his book, *90 Minutes in Heaven*, Don Piper, a Baptist minister, describes an incident after he was involved in a tragic car accident. He was dead for ninety minutes. He came back live. During that time, according to Piper, he was in heaven. He saw beauty. He saw his relatives and loved ones. He describes how his senses were overwhelmed. Heavenly music, incredible beauty, and praise for Lord were everywhere, he said. It filled his heart with the deepest of joy and contentment. Many similar experiences have been recounted by those who have experienced near-death scenarios in their lives. Usually, such experiences are life changing and extraordinarily impactful for those involved. These experiences make most of us wonder, is there something that transcends death? Is there an immortal and immortality?

Anything that is born will eventually die, seems to be a law that our universe lives by, at least according to our perceptual reality. One may go further and generalize that everything in our visible universe is subject to entropic decay or one may say eventual death. However, in most cultures, there is one immortal entity and it is God or divinity. Only God is immortal; it is never born, and it never dies. It would be an incredible story if we could find God. Or just know for sure, is there a God?

## Encounter with Divinity

Surveys after surveys have revealed that a great majority of modern humans believe in God. As high as 90 percent of the population, in many human societies, believes in God. By some estimates, there is more money spent in religious or spiritual pursuits than any other pursuits by human beings. Frankly, this comes as much surprise. Scientific ideology dominates today's human society. With so much progress on the technology front, contribution of science to human survival and standard of living is indisputable. It is difficult to argue with this remarkable progress and the principles that have been behind such progress. Scientific belief rooted in old Newtonian paradigm demands physical material, experimental proof as the eventual truth. Adhering to such guidelines, no one has been able to prove the existence of God. Clearly, according to such form of science, the existence of God would be at best questionable. A nonvisible, nonprovable entity that has supposedly created all, controls all, as we know of, has sure managed to fool some of the best human minds, if it at all exists.

Wait a minute. Isn't God same as digital divine or cosmic mind or informational *whole*? Hasn't the new science based on quantum reality found God in entanglement, nonlocality, and quantum information processing? Isn't God the supreme universal quantum-computing system or digital divine that I have described as *whole* or cosmic mind? Isn't God the perfect mathematical creature, living in the depth of emptiness, a dancer with moves that manipulate every single bit or qubit to all the interconnected bits and qubits and everything in between, an informational Natraj?

The description of *whole* from a computational point of view does come close to describing the divinity as omnipresent, omnipotent, and omniscient entity that has created us and all we see in physical universe and transcends space-time. It probably will hold true on immortality standards as well. Nevertheless, there is a lot more to God than the computational model of digital divinity. Can it describe the God the protector or father figure? Can it describe the personal God that one can communicate with? Can it describe the God that answers prayers? Can it describe God that does miracles? Can it describe God that takes birth to protect weak and kills the evil? Can it

At a deeper level of self realization, I'm
a collection of quantum entities that
simultaneously process information in parallel,
in local, and in nonlocal domain. When I merge
with this reality, I become a silent witness to
myriad of processes happening all in the same
moment of time called *now*, no boundaries
between me and others, a seamless existence, all
coexisting into one. I feel love, empathy, and
profound connectedness and fearlessness.

—Anonymous

describe the God that is always engaged in battle with evil? Besides, most important than perhaps most, can it explain the feeling of divinity by humans that is subjectively experienced through encounters with divinity, the grand feeling of love and oneness?

Clearly different religions have described different types and characteristics of God. The digital rendition of divinity may fall short of fulfilling many of these claims. However, some characteristics emanate naturally from the digital description are widely accepted as the characteristics of God. Most widely recognized is *oneness* with omnipresence, omnipotent, and omniscient. These characteristics at the most basic level represent a union of supreme love that unites all and grand cosmic mind that manages all.

Combining love and logic can only be done in a quantum bit. Where love is equivalent to superpositional state of quantum bit, and binary state is tied to logic. All quantum bits in our universe share one superpositional state of being. This is one God-like characteristics shared by many religious and spiritual descriptions of God. Since this state is also a state where intense parallel computations take place, it is also a state of supreme logic which coexists with eternal love in cosmic mind. It almost appears that we can describe such a God with logic or mathematics of quantum physics as an informational superpositional state of being. Since this state is also the state of emptiness, which occupies most (>99.99999999%) of the universe, it is omnipresent. It is empty because there is no material or energetic reality contained in this state. However, it is also the state of fullness as all quantum computations that run this universe take place in this state. It is also unifying state of grand love since all informational entities contained in this state act as one. Since humans are equipped with love sensors, can we experience this grand unifying state of love?

For human beings, combining the love and logic has always been difficult. We can objectively study logic, but not love. To understand love, we must transcend objectivity and take help of subjective understanding. Here knowing divine becomes a subjective, personal, or emotional experience; either you have it or not. Here is an old story that Prem Rawat or Maharaji shares with seekers of self-knowledge to start them on this path.

Divinity is the *truth* that can be felt; it is instantaneous. Mind is mostly the perception that takes time to
the feeling. I always give priority to *truth* over perception. When I accept truth in my life, I feel *joy*.

—Anonymous

## Doubt or Devotion

There was an owl sitting on a tree that had just woken up. It was dark all around except for the light coming from the stars. Suddenly he saw a swan landed on a branch of the tree right next to him. Swan was a little frustrated and muttering a few words, expressing frustration on what went on throughout the day. Owl, watching quietly, asked him about why he was so frustrated. Swan said, "You know, I started this morning when Sun was just rising. I thought I would reach here by the sunset. Nevertheless, look, it is already so late. I need rest before I start my journey again in the morning."

Owl was puzzled. He asked what Sun is and what is this phenomenon called morning. "You do not know?" Swan was surprised. "Morning is the most beautiful part of the day. Light is everywhere. It comes from Sun. Sun is the most important to all beings on Earth, and no one can live without Sun."

"What Sun! I have been living here for the past ten years, in fact, all my life and I have never seen Sun. When I sleep, it is dark, and when I wake up it is dark as well. Well I like light," said the owl. "I love light coming from stars. Is sunlight similar to the light that comes from stars?"

"Nothing like that," said Swan. "Everything looks different in sunlight. If you like, I could show you the Sun. In fact, you should see the Sun."

"Wow, I would love to. Let me take permission from my parents and others." So the owl went to his parents and asked, "Have you seen anything like Sun? This swan is singing the glory of Sun, and I cannot believe I do not know anything about it."

"You are not alone," the owl's parents remarked. "All our lives we have not known of anything like Sun." Therefore, they went to the elderly of the village and asked the same question. No one knew about Sun. The intellectuals of the village began heavy debate on the existence of Sun.

Swan very surprised said, "You do not have to debate. I can show you the Sun."

"Really! How?"

"All you have to do is to stay awake, and you will see."

"Oh! We cannot do that! It will bring the curse of "darkness" from our God."

One old owl said, "I remember an owl tried it. He never came back."

In this story, we are the owls, the enlightened master is the swan, and of course, Sun is God. Until the doubt turns into devotion, it is not possible to encounter divinity. Moreover, doubt does not go away easily. Roots of doubt are deep. After all, it is the key tool for scientific and objective inquiry that has been highly successful in guiding us to reach and understand our material or even digital reality. The doubt has to play out completely in a subjective mind. When one learns to doubt the doubt itself, a new understanding beyond doubt arises. It is devotion. An ancient path that led seekers to God!

Since ancient times, humans have been divided into ones that claim to have seen or experience God or divine feeling and the ones that have not. Even though different religions and spiritual practices differ in execution, the fundamental nature of what is God or divinelike is similar. Prayer, for example, describes a state of mind, which is a result of being in the company of God, is very common in many religious or spiritual practices.

Ancient Ayurvedic seers spent extensive efforts to develop the knowledge that led them to divinity. Knowing the God, for them was the most supreme of the knowledge. According to them, the following are the paths to know God:

Devotion, Prayers
Action
Self-knowledge
Meditation

Many following paths shown by such knowledge arrived at the divine feeling of grand love. It, however, was subjective; the biggest challenge the ancient seers faced was that God could be felt but could not be fully described. For many, who arrived at that knowing, there were no words or behavior patterns that could describe what was felt. Therefore, many simply became quiet and did not break their silence even in their death, fearing the inadequate oral or written description will corrupt or misguide others to arrive at the true nature of what was felt. Many others tried the description. The books after books could not do justice to description of what God is. Songs after songs could not do justice to what was felt as a true feeling of knowing or dancing with divinity. In the Christian traditions, it was clear that not enough could be said about the glory of God. No matter what or how much was said, it fell short.

The Hebrew word for *glory* is *kabod*; it means weight. In science, it would be the mass of an object of matter. It is the essence of a person or thing. For God, it is who he is, his character and influence. We know that God is love (1 John 4:16); love is God's character and influence. God's glory manifests and reveals his love.

Moses asked God to see his glory. This was God's response, "I will make all my goodness pass before you, and I will proclaim the name of the Lord before you. I will be gracious to whom I will be gracious, and I will have compassion on whom I will have compassion" (Exodus 33:19).

The word *gracious* in Exodus 34:6 means to show favor, mercy, kindness, and forgiveness; longsuffering means to be patient; goodness means to show loving kindness; truth means to be faithful and trustworthy. All of these characteristics are seen as characteristics of love. All the other characteristics of Exodus 34:6 are contained in this compassionate love of God.

In Exodus 34:7, God reveals that he is just. Even justice is a characteristic of love; love must be fair. God is the great equalizer. We know God through experiencing his love, so we may be may be filled with all the fullness of God. The glory of God is the expression and revelation of his love.

A vivid account of the immortal God is attempted in Hindu holy book Bhagavad Gita that means song of God.

Nonlocal self resides within. It is the soul and pure consciousness. In concert with the local self it forms the self.

—Anonymous

In verse number 1 to 3 of chapter 7, Krishna said to Arjuna, "Oh Arjuna! I am going to tell you *Tatwa Gyan*," knowledge or eternal wisdom. It is the knowledge of stuff that makes the universe or existence. Two types of matter create this universe. First makes earth, water, fire, air, sky or universe, mind, intelligence, and even *ego*. It is called form, matter, or *Apara*. The second is vibrations or nonmatter which is formless. It is also known as *Para* or *Chetan* (verse 5). All are created by merging these two. "I am the creator and destructor of these creations where vibrations and matter merge." The description of vibration comes remarkably close to describing the wavelike nature of the quantum universe, the vibration being the entanglement or the wave of guidance.

"There is nothing beyond me. All are part of me. I am the feeling of thinness in Water. I am the light of the Sun and the Moon. I am 'Onkar,' I am manliness of men, I am pure smell on earth, I am the heat in fire, life in all creations and intelligence in intelligent persons. I am the strength of the strong, I am the sexual feeling (Kam), and all other emotions in creations." Krishna further says, "Oh Arjuna! I am *Atman* (soul), and I am in the heart of all creations. I am the start, midpoint, and the end point of all creations. I am mind in *Indriyon* or senses. I am vibrations or *Chetanta* in Creations. I am Om Onkar in words. I am silence, and I am *Tatwa Gyan* or knowledge of everything. There is no creation without me." This would come close to the description that the informational entity *whole* or cosmic mind would provide about itself if it could talk like a human being.

At the deepest level, the experience of meeting God or Maha-Samadhi is accompanied by extraordinary feeling of timelessness and love. A feeling of oneness, with universe, where all dualities merge, a freedom from space and time and entry into grand divine which allows one to move in space-time that is to go anywhere without actually physically going there. The poet Kabir (1398-1448) described the duality of this absolutely extraordinary spiritual realm with extreme clarity. Kabir was illiterate and never wrote anything himself. He merely uttered his feelings to the best of his abilities. In describing God, the art or skills of listening is as important as the art or skills of describing. Even if one could provide a perfect description of the ultimate reality of God, the listener or the recipient of the description may not understand with that perfection. He or she may induce his or her own prejudices to the

It is easy to say that God, in its totality, is everywhere, but the true magic occurs when I understand that he in his totality is inside of me.

Like quantum hologram, complete hologram is part of all points on the hologram. He is in the information regime, the nonlocal regime, beyond space and time.

—Anonymous

interpretation of the description. Therefore, even a self-realized master may not be able to penetrate an imperfect listener.

Since subjectivity is the key element of knowing the God, there is a distinct dividing line between believers and nonbelievers. For believers, the path to know God involves acceptance, love, and trust. Ultimate communion with God is a feeling of eventual joy, oneness, and liberation that is experienced in a deep state of prayer, meditation, and *Samadhi*. For nonbelievers, a proof positive based on objective material and perceptual reality.

One of the pioneers of quantum computing David Deutch said, "The hardware matters!" In this amazing game of existence, the observer or *self* seems to matter. What is observed, how it is observed, what is understood, or what meaning is eventually assigned to the observation seems to matter. The love, kindness, empathy, curiosity, logic, anger, greed, separateness, or oneness is part of the personal reality of *self*. The relationship of *self* to *whole* is also a very subjective to *self*. The relationship on one self to the other *selves* is also subjective in meaning and significance. Is there an experience of oneness buried in all of us that is beyond subjectivity? Is it love that brings the experience of oneness? Is it when heart is full of gratitude, appreciation, love, empathy, and kindness the oneness arises?

At the most fundamental level, the *whole* transcends space-time that is the tiniest and the largest have no meaning. At this level, the cosmic mind is abstract, pure mathematical, quantized, and informational in nature. In its purity, it has no space and time constraints attached to it. It is everywhere and nowhere all at the same time. Information, knowledge, and wisdom all merge into instantaneous knowing or known.

Can we understand our universe from this bottom-up approach where immaterial reality results into a material reality? A reductionist view may imply a yes. However, the answer is no. No matter how accurate our scientific or mathematical models become, a true understanding of comic mind needs a holistic view, involving experiential reality. How is that possible? You may ask. Universe is infinite and transcends outside space and time, all of us; our measuring instruments are all stuck as mere parts of space-time and have only an inside view of the system. You are right. If cosmic mind wanted to stay hidden from its inhabitants, there would be no way for us to know its true

> Divinity exists as part of *your true* reality,
> inside us, gateway to the rest; you can interact
> with it like a real person. You can reduce your
> discontentment, anxiety, and anger by handing
> over your problems to *him*. You can rejoice and
> express gratitude and appreciation for all *he*
> does for you.
>
> —Anonymous

nature. However, it is not so. The cosmic mind is a constant companion and participant of all actions taking place inside this grand mind. In the spirit of a true grand computing system, the cosmic mind incessantly calculates all its aspects, all entities living in, and their interactions at both physical and abstract level at all times. The results of such computations, the wave of guidance, are made available or sent back to all quantum, micro, or macro entities residing in the cosmic mind. Once tuned to this wave of information, it becomes a portal or door to enter or to know the mega computing system, the *whole* or the cosmic mind.

Saints and mystics throughout the world have described their experience of merging with this wave of guidance as the key step to commune with God. Because the wave of guidance is nonlocal information wave, its description had been a challenge. Whatever was truly felt, it was difficult to describe it in terms of normal physical experiences. Most described it as a special sound. Most sounds heard by us have a source that produces them. Most of these sounds have intensities that diminish with distance from the source. This sound, these masters claim, comes from no source. It has no origin, and it never ends. Its intensity never lessens and is independent of distance, time, or location. All masters, from all different era in time and location, have the same experience of hearing this sound. Ayurvedic spiritual masters have described this as a sound, an eternal sound. The Hindu word *om*, really represents this sound. The sound of *om* comes from the experience of this wave of guidance by the spiritual masters. In their deep experience of Samadhi, the sound of *om* is heard. Christian and Jewish masters heard this as "amen." Hindus consider *om* to be the supreme divine source, which created this universe.

For all the unconscious entities, the wave of guidance directs the entity to a preferred state of being. All entities, by default, follow the wave of guidance except for the ones with consciousness and free will such as us humans. We may choose to ignore messages from cosmic mind at will and follow the cue from our local rational mind. Once tuned, the wave of guidance projects a persona of personal guide or God to such entities. Such experiences are commonly reported by followers of many religions. The connection to this wave of guidance results in the description of the cosmic mind as the omniscient, omnipresent, and omnipotent God illusive to many and a constant companion to others. This waver of guidance and our ability to

At the deepest level of my existence there is this profound oneness, profound fullness or contentment or completeness. The witness is lost in the rest. Eternity is arrived. The feeling of immortality, timelessness, spacelessness, and weightlessness become the nature.

—Anonymous

tune to it are the root of all religions, spirituality, and mystic experiences, in my view.

While trekking down the road to digital divine may paint a mental image of what may potentially be divine, the perception of true divinity can only come as a feeling. It needs a subjective *self* with a knowing of its true quantum nature. The subjective experience of "deep joy from within" is at the core of this reality. It is nothing short of enlightenment. May all arrive and get to drink this incredibly sweet and intoxicating divine sap.

For more Information and free downloads, visit

*http://www.drhemantgupta.com*
or
*http://www.roadtodigitaldivine.com*

# Notes and Credits

Chapter 1

Selected Quotations, http://www.its.caltech.edu/~sjordan/quotations.html.
http://www.quotes.ubr.com/subject-quotes/d/divinity-quotes.aspx.

Seth Lloyd, *Programming the Universe.*

Chapter 2

WikiSD, "System," http://www.systemdynamics.org/wiki/index.php/
System.

Chapter 3

*Discover Magazine,* "The Biggest Chill: Subatomic Particles," http://
discovermagazine.com/1993/feb/thebiggestchill174.

"BEC: Temperature and Absolute Zero," http://www.colorado.edu/
physics/2000/bec/temperature.html.

"Emptiness Is Form," http://www.thebigview.com/buddhism/emptiness.
html.

Chapter 4

Reuters, "Vatican Should Learn from Galileo Mess, Prelate Says," http://
www.reuters.com/article/idUSTRE5614DL20090702.

"Introducing the Universe," http://www.tricitiesnet.com/donsastronomy/intro.html.

"Office Post Advanced," http://www.humphreys.edu/faculty/jdecosta/Jim/ADM270/wk4lecture.htm.

"Nanotechnology," http://nanotechnic.blogspot.com.

"Opportunities in Atomic, Molecular, and Optical Physics," http://www.fen.bilkent.edu.tr/~bulutay/438-538/Compilation-AMO-2010-Report.pdf.

"Teacher's Guide to Atomic Scale Microscopy and Powers of Ten," http://chem.lapeer.org/PhysicsDocs/Goals2000/Lesson1.html.

"KryssTal: The Scale of the Universe," http://www.krysstal.com/scale.html.

"Atomic Theory I," http://www.visionlearning.com/library/module_viewer.php?mid=50.

"Elements of Chemistry: Atoms: The Building Blocks of Matter," http://school.discoveryeducation.com/lessonplans/programs/ec_atoms/.

"Nanotechnology," http://nanotechnic.blogspot.com/.

"Opportunities in Atomic, Molecular, and Optical Physics," http://www.fen.bilkent.edu.tr/~bulutay/438-538/Compilation-AMO-2010-Report.pdf.

Chapter 5

"Average American Consumes 34 Gigabytes Daily," http://www.dslreports.com/shownews/Average-American-Consumes-34-Gigabytes-Daily-105910.

"The Computational Universe," http://www.edge.org/3rd_culture/lloyd2/lloyd2_print.html.

"Evolutionary Quantum Computation," http://www.goertzel.org/dynapsyc/1997/Qc.html.

"Imaging Quantum Entanglement," http://www.sciencedaily.com/releases/2007/09/070921112416.htm.

"Secret of Instantaneous Communication and Reality of Nature," http://www.unexplainable.net/artman/publish/article_3647.shtml.

Antoine Henri Becquerel, http://stwww.weizmann.ac.il/g-junior/matmon-new/common_tools/scientists/bell.htm.

"Overview of the Derivations," http://www.lecb.ncifcrf.gov/~toms/papers/edmm/latex/node3.html.

"Quantum Algorithm Solves Systems of Linear Equations," http://www.scientificcomputing.com/news-DA-Quantum-Algorithm-Solves-Systems-of-Linear-Equations-101309.aspx.

Chapter 6

TED Blog, "Henry Markram at TEDGlobal 2009," http://blog.ted.com/2009/07/henry_markram_a.php

Chapter 7 http://cfapps.venturacountystar.com/metrolink/person.cfm?personID=38.

R&D Mag, "Protein Motor Springs to Action," http://rdmag.com/News/2009/11/Life-Science-Protein-motor-springs-to-action/.

"Bodywork and Neuropeptides," http://www.healtouch.com/csft/bodywork.html.

"Information, Science, and Biology," http://creation.com/information-science-and-biology.

"The Wisdom of Your Cells," http://www.energetic-medicine.net/bioenergetic-articles/articles/33/1/The-Wisdom-of-Your-Cells/Page1.html.

Candace Beebe Pert, *Molecules of Emotion:* The Science Between Mind-Body Medicine (Scribner, 1999).

Candace Beebe Pert, *Everything You Need to Know to Feel Go(o)d*, with Nancy Marriott (Hay House, Inc., 2006).

Deepak Chopra, *Return of the Rishi* (1988).

Deepak Chopra, *Quantum Healing: Exploring the Frontiers of Mind/Body Medicine* (1989).

Deepak Chopra, *The Path to Love: Renewing the Power of Spirit in Your Life* (1997).

Deepak Chopra, *The Seven Spiritual Laws for Parents: Guiding Your Children to Success and Fulfillment* (1997).

Deepak Chopra, *Everyday Immortality: A Concise Course in Spiritual Transformation* (1999).

Deepak Chopra, *How to Know God: The Soul's Journey into the Mystery of Mysteries* (2000).

Deepak Chopra, *The Spontaneous Fulfillment of Desire: Harnessing the Infinite Power of Coincidence* (2003).

Deepak Chopra, *Synchrodestiny: Harnessing the Infinite Power of Coincidence to Create Miracles* (2003).

Deepak Chopra, *Life After Death: The Burden of Proof* (2006).

Deepak Chopra, *Kama Sutra: Including the Seven Spiritual Laws of Love* (2006).

Deepak Chopra, Buddha: A Story of Enlightenment (2007).

Chapter 8

"Wired 10.12: God Is the Machine," http://www.wired.com/wired/archive/10.12/holytech_pr.html.

Boing Boing, November 19, 2002 archives, http://www.boingboing.net/2002/11/19/.

Bhagavad Gita.

Connectedness, http://www.plim.org/Connectedness.htm.

Grand Universe of Primary Consciousness Quantum Essays, http://primordality.com/quantum_theory.htm.

Chapter 9

Don Piper, *90 Minutes in Heaven*, www.90minutesinheaven.com.

"Has Science Found God in Nonlocal Reality?" http://www.plim.org/nonlocal.htm.

"Hinduism and Quantum Physics," http://www.hinduism.co.za/hinduism_quantum.htm.

"The Glory of God," http://www.seekgod.org/message/gloryofgod.html.

Srimad Bhagwat Geeta, chapter 7, http://sites.srimadbhagwatgeeta.com/chapter-7/.

Kabir Poems, http://midiagols.com.br/africa2010/quadro.php?help=Kabir+Poems.

"Fingerprints of God: The Search for the Science of Spirituality," National Public Radio correspondent, Barbara Bradley Hagerty, barbarabradleyhagerty.com/.

Sri Chinmoy, "Love, Devotion and Surrender," http://www.srichinmoy.org/polski/resources/library/talks/inner_qualities/love_devotion/.

BBC News, "Simulated Brain Closer," http://news.bbc.co.uk/2/hi/science/nature/8012496.stm.

Prem Rawat or Maharaji, http://www.tprf.org/.

Osho (Bhagwan Shree Rajneesh), www.osho.com

Throughout writing of this book, various quotes simply came and stayed in my mind and I was able to write them down. These are the Quotes designated from Anonymous as I do not know the origin of these.

All images used in this book have been obtained from Wikimedia Commons which in turn cites images source as Flicker's The Commons.

# Informational
# Nature of Being

**BOOK II (Summary)**
**JOY FROM DEEP WITHIN**
**TRUE NATURE OF YOUR QUANTUM SELF**

What do we all humans have in common?

We are all incorrigible seekers of bliss or joy. Our minds are wired to be happy. At least that is what our modern scientific studies contend. Yet most of us know we are not joyful. In fact, we struggle most of time to be and stay happy. What has gone wrong?

I have attempted to answer this question in this book. It has been my own journey to arrive at lasting happiness and joy. It builds on the informational, computational foundation of our universe developed in the first book of this series, *Road to Digital Divine*. It combines latest science of mind and matter with spirituality, putting forth a new concept of our self, the quantum self rooted in the informational nature of our being.

Most of us are familiar with our physical self that we see in the mirror. It, however, does not define us completely. What is critical for us to know is our informational *self*. Knowing the true nature of this self and acting accordingly is essential for us humans to achieve good emotional health and realize lasting joy in our lives. In fact, not knowing or ignoring the messages from this essential nature of our *self* is the leading reason for suffering in our human society.

I paint this self as a tale of three minds: emotional, rational, and cosmic. The three minds computationally synchronized lead to the birth of informational self. This self is a computationally astute structure. It computes using two modes of computation. Each of these modes lead to two extreme personalities. One, binary computation, which leads to a nature of self that courts "I, me, and mine" tendencies. I have called this as our binary self. And the other, quantum computation, which leads to a nature of self with "us, we, and ours" tendencies. I call this as our quantum self.

Both these nature of self are quite familiar in today's society. The binary information processing leads to egoic entity, which is present in most of us. It dominates today's human society. It is responsible for incredible progress that humans have made as a surviving species, but it is also responsible for most of the sufferings that modern humans face today. The quantum self has saintlike nature. It feels love, empathy, and oneness with others. It is truthful and always stays in the company of divinity. It is responsible for widespread altruism in nature and in humans.

With two selves of very different nature in one body, modern humans have learned the meaning of the word "suffering." Which self will win? Which is our true self? Understanding of this fact is not trivial. In fact, it is nothing short of enlightenment as I explain how the understanding of our true self can lead one on to this path. It can make a profound change in one's perspective. The joy pouring from deep within is at the root of this reality. For more information please visit author's websites: *http://www.drhemantgupta.com* and *http://www.roadtodigitaldivine.com*.

# *Index*

## K

Kabir (poet), 187
Krishna (Hindu divinity), 164, 187

## L

Landauer's principle, 91
Leibniz, Gottfried Wilhelm, 126
levels of existence
  experiential, 25
  mathematical or informational, 25
  mental, 25
  physical, 25
life, assumed origins of, 135
living and nonliving, distinction
    between, 115
Lloyd, Seth, 99, 101, 195
  *Programming the Universe*, 23
local group, 59

## M

machine, classical definition of, 32
Maha-Samadhi, 187. *See also* God
Markram, Henry, 118, 197
matter, 32-33, 42, 63
  atomic theory of, 66
Maxwell, J. C., 91
Maxwell's demon, 90-91
Milky Way, 57, 59
mind, 162
molecules, 63, 65
*Molecules of Emotions* (Pert), 147

## N

neocortex
  continued evolution of, 125
  repertoire of functions in, 119
nerve cells, 121
nervous system, 122
Neural connectivity, 119, 121
neuropeptide, 151
neutrons, 44, 67, 73
Newtonian physics. *See* classical
    physics
nihilism, 45
*90 Minutes in Heaven* (Piper), 178,
    199
nonlocal communication, 97
nucleus, 67, 137

## O

observational views
  bottom-up, 34
  top-down, 35
Oxytocin, 147

## P

particles, 69, 71
  quantum, 44
Pascal, Blaise, 123
*paticcasamuppada*, 126
peptides, 140, 145, 147, 149, 151
Pert, Candice, 147, 149, 151
photons, 67, 70
physical reality, 163, 167, 169-70, 173
Piper, Don, 178, 199

www.ingramcontent.com/pod-product-compliance
Lightning Source LLC
Chambersburg PA
CBHW032002170526
45157CB00002B/502